Engineering and Technology Management
Tools and Applications

For a listing of recent titles in the *Artech House Technology Management and Professional Development Library*, turn to the back of this book.

Engineering and Technology Management Tools and Applications

B. S. Dhillon

Artech House
Boston • London
www.artechhouse.com

Library of Congress Cataloging-in-Publication Data
Dhillon, B. S.
    Engineering and technology management tools and applications/B.S. Dhillon.
        p.   cm. — (Artech House technology management and professional
    development library)
    Includes bibliographical references and index.
    ISBN 1-58053-265-9 (alk. paper)
        1. Engineering—Management. 2. Technology—Management. I. Title. II. Series.
TA190 .D47   2002
620'.0068—dc21   2002074491

British Library Cataloguing in Publication Data
Dhillon, B. S.
    Engineering and technology management tools and applications.
    (Artech House technology management and professional development library)
    1. Engineering—Management   2. Technology—Management
    I. Title
    620' . 0068

    ISBN 1-58053-265-9

Cover design by Igor Valdman

© 2002 ARTECH HOUSE, INC.
685 Canton Street
Norwood, MA 02062

International Standard Book Number: 1-58053-265-9
Library of Congress Catalog Card Number: 2002074491

10 9 8 7 6 5 4 3 2 1

Affectionately dedicated to my daughter, Jasmine

Contents

Preface

Today engineering and technology have become important factors in global competitiveness. Under a fierce competitive environment, engineering and technology managers are forced to achieve marketable results by focusing on areas such as speed, quality, and cost. This can only be achieved through effective planning, organization, and integration of rather complex multidisciplinary activities across functional lines, a great amount of people skills, and effective integration of new knowledge. More specifically, it basically means that an effective knowledge of engineering and technology management is absolutely essential for the survival of business and the maintenance of national prosperity in today's fiercely competitive global economy.

Today, there are a large number of books available on engineering management and technology management, but—to the best of my knowledge—there is none that emphasizes tools and applications such as concurrent engineering, reengineering, value engineering, reverse engineering, *total quality management* (TQM), configuration management, software engineering management, and *information technology* (IT) management. Knowledge in these areas is considered essential for engineers and other technical professionals to manage today's business activities effectively.

Currently, information on these areas is mostly available either in specialized books or in articles but not in a single volume. Thus, the main purpose of this book is to satisfy this vital need. This book generally focuses on the structure of concepts rather than the minute details. The sources of much of the material are given in references at the end of each chapter, which will be useful to readers who desire to delve deeper into specific areas. The book is composed of 17 chapters.

Chapter 1 presents various introductory aspects of engineering and technology management, including history of management, management characteristics and functions, engineering and technology management challenges and skill needs, useful information on engineering and technology management, and scope of the book. Chapter 2 is devoted to organization and the human element and covers topics such as methods of organization, span of control, functions of engineering departments, qualities and activities of engineering managers, and committees and staff meetings.

Chapter 3 presents important tools for making effective engineering and technology management decisions, including decision trees, optimization techniques, discounted cash flow analysis, learning curve analysis, depreciation analysis, fault tree analysis, and forecasting methods. Chapter 4 covers various topics concerning project selection and management. Some of these topics are project selection methods and models, project management techniques, and project manager's responsibilities, qualifications, selection, and reporting.

Chapter 5 is devoted to the management of engineering design and product costing and includes topics such as design types and approaches, engineering design manpower, design reviews and design review team, reasons for product costing, product life cycle costing, and new product pricing. Chapter 6 presents the management of proposals and contracts. This chapter covers topics such as technical proposal types, proposal development procedure, proposal elements, engineering specifications, contract types, contractor selection factors, the contract negotiation approach, and mathematical models for contracting.

Chapters 7 and 8 are devoted to creativity and innovation and concurrent engineering, respectively. Some of the topics covered in Chapter 7 are creativity methods, the creative problem-solving process, and barriers to creative thinking. Chapter 8 includes topics such as concurrent engineering objectives and concepts, the concurrent engineering team, and the concurrent engineering process related methodologies and techniques.

Chapter 9 presents various aspects of value engineering ranging from value engineering objectives to value engineering phases and techniques. Reverse engineering is presented in Chapter 10. This chapter covers topics such as reverse engineering objectives, reverse engineering methods, software reverse engineering, and the reverse engineering team.

Chapter 11 is devoted to configuration management and covers topics such as reasons for having a configuration management system, configuration management plan and disciplines, the configuration management organization, and software configuration management. Chapter 12 presents various aspects concerning TQM, including TQM principles and elements, TQM methods and techniques, and barriers to TQM success.

Chapter 13 presents maintenance management. This chapter covers topics such as maintenance department functions and organization, a maintenance management approach, effective maintenance management elements, and performance measurement indexes. Chapter 14 is devoted to warranties, ethics, and legal factors and presents topics such as reasons for the warranty needs, warranty types, warranty components and management, need for ethics, general guidelines for ethical behavior, product liability and patents, and copyrights and trademarks.

Chapters 15 and 16 are devoted to two important topics: reengineering and IT management, respectively. Some of the topics covered in Chapter 15 are reengineering facts and figures, the reengineering process, reengineering tools, the reengineering team and manpower, and reengineering guidelines. Chapter 16 includes topics such as IT management–related facts and figures, IT manpower, network management, the client-server system, and human factors in information systems.

Chapter 17 presents various aspects concerning software engineering management. Some of the topics covered are software facts and figures, software engineering project management, useful models for software engineering management, software engineer's functions and skills, and software engineering management standards.

The book will be useful to many individuals including senior level undergraduates and graduate students in engineering or technology management, industrial engineering, manufacturing engineering, production engineering, engineering in general, college and university level teachers, engineering or technology managers, and professional and nonprofessional engineers and technologists.

1

Introduction

The master works of ancient Persian, Greco, Roman, Chinese, and Scythian civilizations all involved engineering to varying degrees, and the individuals responsible for those discoveries and enterprises were in the every respect engineers [1]. The word "engineer" is derived from the Latin *ingenium*, which literally means natural talent or capacity as well as "a clever invention" [2]. Nonetheless, in the modern terms the words "technology" and "engineering" may simply be described as the knowledge to manipulate nature to produce products (consumer, industrial, and commercial), energy, and services; and the understanding of the manipulation process that seeks to satisfy human social and economic needs and aspirations, respectively. Both engineering and technology interface with mathematics, physical sciences, social sciences, and humanities. An effective understanding of the former is absolutely necessary to translate innovative ideas into reality while due regard for the latter is critical for business success.

Engineering and technology management is basically concerned with managing engineering and technologies to achieve business objectives or goals. It requires skills in understanding technology and engineering in addition to managing business activities of organizations.

In today's fiercely competitive global economy, an effective management of engineering and technology is crucial for the survival of business and the maintenance of national prosperity. This chapter presents topics directly or indirectly concerned with the introductory aspects of engineering and technology management.

1

1.1 History of Management

Although, the history of the word *manage* may be traced back to the Latin word *manus* meaning "hand," the history of modern management begins with the Industrial Revolution. The Industrial Revolution was the result of the development of the steam engine by James Watt in the eighteenth century in Great Britain. In 1830, shortly after the introduction of the engine in the United States, Colonel John Stevens, the father of American engineering, built the first 23-mile-long railroad. By 1850, the total railroad track increased to 9,000 miles extending as far west as Ohio. Another factor that played an instrumental role during the Industrial Revolution was the development of the telegraph by F. B. Morse. The first experimental telegraph line was built in 1844, and by 1860 there was a total of 50,000 miles of telegraph line in the United States [2].

Before 1835, there were only 36 firms in the United States that employed more than 250 workers [3]. During the last decade of the nineteenth century, persons such as John D. Rockefeller, Andrew Carnegie, and Cornelius Vanderbilt took advantage of railroads and telegraph lines to build big corporations employing thousands of people. In turn, this led to the need of a systematic approach to management.

Two engineers who may be called the fathers of modern management were Frederick W. Taylor (1856–1915) and Henri Fayol (1841–1925). Frederick W. Taylor was born in Philadelphia; although accepted into Harvard University, he served a 4-year apprenticeship as a machinist [2]. In 1878, he joined Midvale Steel Company, and at the age of 28, in 1884, he became chief engineer. In 1906, Taylor became president of the American Society of Mechanical Engineers, and his basic views concerning management were finding the most appropriate method for performing a job and assigning the right person for each job.

Henri Fayol graduated in 1860 from the National School of Mines at Saint-Etienne, France, and outlined 14 principles of management, including division of work, discipline, line of authority, initiative, order, and centralization. In 1916, he published a book entitled *Administration Industrielle et Generale* covering most of his thoughts on management [4, 5]. The book was translated into English twice: in 1930 by J. A. Coubrogh and in 1949 by C. Storrs [6].

In 1911, the first ever conference on the topic of scientific management was held, and during the period from 1912–1936 various professional societies concerned directly or indirectly with the promotion of management were formed. For example, in 1912, 1917, 1923, and 1936, the Society to

Promote the Science of Management, the Society of Industrial Engineers, the American Management Association, and the Society for the Advancement of Management were established, respectively. By 1925, most engineering schools in the United States were offering some kinds of courses on management [7].

In 1924, a study on various aspects of human relations (e.g., investigating the effects of varying illumination, length of workday, and rest periods on productivity) was initiated by the National Research Council of the National Academy of Sciences at the Hawthorne Plant of Western Electric in the state of Illinois. The findings of this study also played an important role in the development of the management field. Since those days, thousands of individuals have contributed to the management field. A vast number of publications in the form of books, conference proceedings, and journal articles have appeared, and thousands of university-level institutions award undergraduate and graduate degrees in various aspects of management around the world. Furthermore, the field of management has branched out into many specialized areas, and engineering and technology management is one of those areas.

1.2 Terms and Definitions

This section presents selective terms and definitions directly or indirectly concerned with engineering and technology management [6, 8–12].

- *Management.* This is a process of work involving guiding a group of individuals to achieve defined organizational goals.
- *Engineering and technology management.* This is concerned with managing engineering and technologies to achieve business objectives, and it requires skills in understanding technology and engineering in addition to managing business activities of organizations.
- *Project managing.* This is the process of managing the project or the administrative and technical lead people of a project team.
- *System management.* This is the process of planning, organizing, directing, controlling, and coordinating joint efforts to achieve system program goals.
- *Team management.* This is the organizational and functional approach to accomplish compatibility within an organization.
- *Contractor.* This is a company that engages in a contract.

- *Design review.* This is an administrative and technical control exercised to bring to each design the expertise of individuals intimately familiar with the complete designs or segments of the complete designs.

- *Engineering report.* This is a document that provides project clarification on a detailed technical level.

- *Functional organization.* This is an organization engaged in one general function (e.g., marketing, engineering, or manufacturing).

- *Policies.* This is a basic set of project or corporate guidelines to manage organizational resources and attain goals.

- *Problem statement.* This is the documentation for stating and clarifying the problem.

- *Specifications.* These are performance, size, weight, environmental, and other requirements that a deliverable item must satisfy.

- *Verification.* This is an assurance that the resource and time estimates match the overall goals.

- *Scope statement.* This is a document describing the magnitude and depth of the project, including global results.

- *Manpower schedule.* This is the manpower requirements established to satisfy the time schedule.

- *Fiscal management.* This is the management of the flow of funds received from the customer and needed in project operations.

- *Management by objectives.* This is a participative goal-setting process that allows management to construct and communicate the organization's objectives to each employee.

- *Project manual.* This is a document that contains all written contract papers excluding drawings, and it may also include such items as sample forms, general and supplemental conditions, and bidding papers.

- *Schematic.* This is a graphic illustration depicting construction and operation principles without accurate mechanical representations.

- *Engineering change.* This is a revision to a parts list, bill of materials, or drawing produced and authorized by the engineering organization/department and normally identified by a control number.

- *Technology base.* This is a term referring to efforts that contribute technical capabilities and scientific knowledge to the effectiveness of the system and organization.

1.3 Management Characteristics and Functions and Traditional Management Versus Modern Management

There are many characteristics of management including the ones listed next [6, 13].

- Management is intangible.

- Management has a purpose because it is practiced to accomplish a specific goal.

- Management is an activity, not a person or group of people. This activity is carried out by various people (e.g., supervisors, managers, and executives). All in all, just like any other activity, it can also be studied and skills in its application can be acquired.

- Management requires certain knowledge, skill, and practice for its effective use.

- Management is aided, but not replaced, by computers. More specifically, computers can widen a manager's vision and sharpen his or her insight by providing appropriate information for important decisions and facilitating the application of quantitative management-related tools.

- Usually, the practice of management is associated with the efforts of a group of individuals.

- The people who practice management are not necessarily the same as the owners.

- Management is a very important means to make things happen.

- Management is an excellent means for exerting a real impact upon human day-to-day life. More specifically, a manager can be a key player in bringing about vision, hope, achievement, and action for the better things of life.

- Management is accomplished by, with, and through others' efforts.

There are five fundamental functions of management, as shown in Figure 1.1 [14]. *Controlling* includes activities such as performance monitoring, comparing actual performance to set standards, and taking corrective actions. *Planning* is concerned with activities such as establishing goals, determining rules and procedures, forecasting, and scheduling. *Staffing* is concerned with activities such as determining the need for manpower, establishing standards for measuring performance of employees, selecting employees, and hiring and training employees. *Organizing* involves activities such as grouping and assigning jobs or tasks and delegating authority to subordinates. Finally, *controlling* is concerned with activities such as maintaining morale, describing set goals to employees, rewarding employees, and guiding employees to meet set performance standards.

With the passage of time, many differences between traditional and modern forms of management have occurred. Table 1.1 presents important comparisons of the two management systems [9].

1.4 Engineering and Technology Management Challenges and Skill Requirements

In today's global economy, the engineering and technology environment has become more challenging than ever before. Managers in this arena must

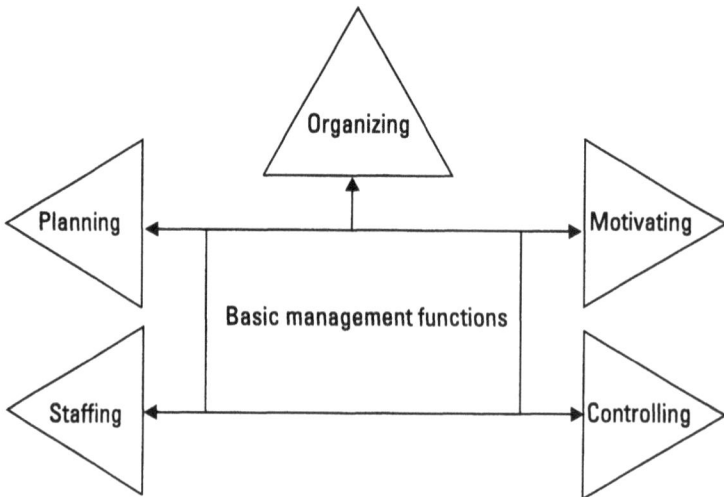

Figure 1.1 Fundamental functions of management.

Table 1.1
Traditional Management Versus Modern Management

Number	Traditional Management (Stable Environment)	Modern Management (Dynamic Environment)
1	Clear-cut single objectives	Multivalued objectives
2	Autocratic leadership	Participative and democratic leadership
3	Authority by position	Well-earned authority
4	Economic-based incentives	Economic- and intrinsic-based rewards
5	Clear-cut tasks	Multidisciplinary tasks
6	Emphasis on individuality	Emphasis on team
7	Management-based controls	Commitment and self-actuating-inclined behavior
8	Individuals trained to fit organization or company	Individuals shape the organization and its future
9	Stability inclined	Change and growth inclined
10	Conflict considered undesirable	Conflict considered unavoidable
11	Strict segmentation of labor force	Transparent multigroup dependencies

operate in a multidisciplinary environment with effective dealings with a variety of interfaces. Moreover, engineering and technology managers have to cope with constant and rapidly changing areas such as technology, markets, socioeconomic factors, and regulations and controls. Nonetheless, some of the important challenges of engineering and technology management are as follows [9].

Changing technology. This is concerned with effective utilization of existing technology and future technological advances, particularly in areas such as design, manufacturing, and service. The impact areas of changing technology include forecasting, flexible planning, risks, market success, opportunities, and profits.

Limited available resources. Here, managers are faced with factors such as tight budgets, resource sharing, limited availability of specially skilled personnel, and shifting priorities. Thus, the impact areas of limited resources include priorities, conflict, creativity, planning, and power struggles.

Task complexity.   Engineering- and technology-related tasks are far more complex and multidisciplinary than their nontechnical counterparts. For example, the design and development of an engineering system require the integration and coordination of enormous activities and subsystems involving people from various functional units with different objectives, desires, and skills. Thus, the impact areas of task complexity include skill requirements, multifunctional direction, work force, organizational design, risks, and creativity.

Multifunctional team-building.   Most of the engineering activities are too multidisciplinary to be structured along functional lines. They involve factors such as broad organizational involvement, multifunctional task integration, and teamwork in all areas. Furthermore, building a good engineering team requires a wide spectrum of management skills to identify, commit, and integrate numerous people and task groups from functional support organizations into an effective single management unit. Thus, some of the impact areas of multifunctional team building are motivation, leadership, organizational design, decision making, commitments, and power struggles.

Date-driven schedules.   Meeting deadlines are realities in any business, but in engineering managers must find alternate solutions to technical problems without sacrificing economy, performance, safety, and the quality of their creations to meet the deadlines. Thus, the impact areas of end-date driven schedules include risks, make-buy decisions, creativity, conflict, and quality.

Resource competition.   Factors such as complex organizational relations, external pressures for efficiency, and market performance often result in intense competition over available resources (e.g., budgets, people, and facilities). This requires engineering managers to conduct skillful planning and resource negotiations. Thus, the impact areas of resource competition include planning, top management control, conflict, commitment, budgets, leadership, and priorities.

Uncertainty and risks.   Factors such as on-budget performance, increasing technological complexities, competitive pressures to complete projects on time, and an uncertain and unpredictable business environment contribute to the uncertainties and risks associated with managing engineering activities. Thus, the impact areas of uncertainty and risks include engineering success, dynamic planning, dynamic leadership, schedules, and priorities.

Creativity and innovation. The winning edge of any engineering organization is its innovative capacity because it directly influences important features such as quality, economy, serviceability, and value of an item or a product. An engineering manager must build environments where people work creatively and innovatively. The main impact areas of innovation and creativity are team building, leadership, power plays, market success, conflict, and profits.

Limited rewards. As traditional rewards such as increments in salary, bonuses, and promotions become scarce, an engineering manager must create rewards based on accomplishments recognition, professional development, work challenges, and freedom. This requires from a manager the integration of various skills into a single homogeneous management system. Thus, the impact areas of limited rewards are turnover, leadership, and motivation.

Managing engineering and technology functions effectively in today's environment requires specific skills in three basic areas: leadership, technology, and administration [9, 15, 16]. The elements of the leadership skill include defining clear objectives, motivating people, building multidisciplinary teams, managing in an unstructured work environment, gaining top management support and commitment, effectively communicating (written and oral), possessing a clear understanding of professional needs, managing conflicts, providing assistance in problem solving, directing clearly, understanding the organization, building priority images, maintaining credibility, and creating personnel involvement at all levels.

Some of the components of the technical skill are managing technology, communicating effectively with technical people, understanding clearly the available engineering tools and support methods, uniting the technical team, understanding clearly technology and its trends, maintaining technical credibility, understanding the market and product applications, facilitating tradeoffs, and integrating technical, business, and human goals.

The components of the administrative skill include planning and organizing multifunctional programs, estimating and negotiating resources, scheduling multidisciplinary activities, delegating activities effectively, measuring work performance and progress, attracting and holding quality personnel, understanding clearly policies and operating procedures, managing changes, and working with other organizations.

It is to be noted that some of the skill components of these three categories may overlap.

1.5 Useful Information on Engineering and Technology Management

In recent years, the field of engineering and technology management has been gaining importance in various sectors including industry and government. Today, there are a vast number of publications available in the form of books, technical papers, and conference proceedings on engineering and technology management. In addition, many professional societies and journals are directly or indirectly concerned with the subject of engineering and technology management.

This section presents selective but useful books, professional societies, and journals directly or indirectly concerned with engineering and technology management.

1.5.1 Books

Some of the books directly or indirectly concerned with engineering and technology management are as follows:

- Babcock, D. L., *Managing Engineering and Technology*, Upper Saddle River, NJ: Prentice Hall, 1996.

- Badaway, M. K., *Developing Managerial Skills in Engineers and Scientists*, New York: Van Nostrand Reinhold, 1995.

- Cronstedt, V., *Engineering Management and Administration*, New York: McGraw-Hill, 1961.

- Dhillon, B. S., *Engineering Management*, Lancaster, PA: Technomic Publishing Company, 1987.

- Dorf, R. (ed.), *The Technology Management Handbook*, Boca Raton, FL: CRC Press LCC, 1999.

- Gaynor, G. H. (ed.), *Handbook of Technology Management*, New York: McGraw-Hill, 1996.

- Hicks, T. G., *Successful Engineering Management*, New York: McGraw-Hill, 1966.

- Khalil, T. M., *Management of Technology*, New York: McGraw-Hill, 2000.

- Lanigan, M., *Engineers in Business*, Reading, MA: Addison-Wesley, 1992.

- Mazda, F. F., *Engineering Management,* Reading, MA: Addison-Wesley, 1998.
- Shainis, M. J., *Engineering Management,* Columbus, OH: Battelle Press, 1995.
- Shannon, R. E., *Engineering Management,* New York: John Wiley and Sons, 1980.
- Ullman, J. E. (ed.), *Handbook of Engineering Management,* New York: John Wiley and Sons, 1986.

1.5.2 Professional Organizations

There are many professional organizations that are directly or indirectly concerned with engineering and technology management. Some of these organizations are as follows [1, 6]:

- American Association of Cost Engineers;
- American Management Association;
- American Society for Engineering Management;
- Association of Consulting Management Engineers, Inc.;
- Institute of Electrical and Electronics Engineers;
- Institute of Industrial Engineers;
- Institute of Management Sciences;
- Society of American Value Engineers.

1.5.3 Journals

Most of the above listed societies publish journals directly or indirectly concerned with engineering and technology management. Nonetheless, some of the journals that publish the latest findings on various aspects of engineering and technology management are as follows:

- *California Management Review;*
- *Engineering Management Journal;*
- *Harvard Business Review;*
- *IEEE Engineering Management Review;*
- *IEEE Transactions on Engineering Management;*
- *Journal of Construction Engineering and Management;*

- *Journal of Engineering and Technology Management;*
- *Journal of Information Technology Management;*
- *Sloan Management Review;*
- *The Wall Street Journal.*

1.6 Scope of the Text

In today's competitive and global economy, an effective management of engineering and technology is very important for the survival of a business. Although a significant number of books on engineering management and technology management are available, none emphasizes modern concepts and applications such as concurrent engineering, value management, configuration management, reverse engineering, reengineering, information technology management, and software engineering management. Knowledge in these areas is considered essential for engineers and other technical professionals to manage today's business activities effectively.

Currently, information on these topics is largely available either in specialized texts or in various articles, but not in a single volume. This book is written to satisfy this vital need. The source of most of the material presented is referenced at the end of each chapter. This will be useful to readers if they desire to delve deeper into a particular area.

This book will be useful to many individuals including senior-level undergraduates and graduate students in engineering/technology management and related areas, college and university teachers, students and instructors of short courses in engineering/technology management, engineering/technology managers, and professional and nonprofessional engineers and technologists.

1.7 Problems

1. Write an essay on the history of the management movement.
2. Define the following terms:

   - Management;
   - Project management;
   - Functional organization.

3. What are the typical characteristics of management?
4. Discuss the basic functions of management.

5. Discuss the important challenges associated with engineering and technology management.

6. Make a comparison between traditional management and modern management.

References

[1] Weinert, D. G., "The Structure of Engineering," in *Handbook of Engineering Management*, J. E. Ullmann, D. A. Christman, and B. Holtje (eds.), New York: John Wiley and Sons, 1986, pp. 13–34.

[2] Babcock, D. L., *Managing Engineering and Technology*, Englewood Cliffs, NJ: Prentice Hall, 1991.

[3] Wren, D. A., *The Evolution of Management Thought*, New York: John Wiley and Sons, 1987.

[4] Fayol, H., *Administration Industrielle et Generale*, Paris: The Society de L'Industrie Minerale, 1916.

[5] Fayol, H., *General and Industrial Management*, London: Sir Isaac Pitman and Sons, 1949.

[6] Dhillon, B. S., *Engineering Management*, Lancaster, PA: Technomic Publishing Company, 1987.

[7] Mee, J. F., "Management Teaching in Historical Perspective," *The Southern Journal of Business*, Vol. 7, May 1972, pp. 21–25.

[8] Banki, I. S., *Dictionary of Administration and Management*, Los Angeles, CA: Systems Research Institute, 1981.

[9] Thamhain, H. J., *Engineering Management*, New York: John Wiley and Sons, 1992.

[10] Fiddler, D. W. (ed.), *Business Terms, Phrases, and Abbreviations*, London: Pitman Publishers, 1971.

[11] Rue, L. W., and L. L. Byars, *Management: Theory and Application*, Chicago, IL: Irwin, 1980.

[12] McKenna, T., and R. Oliverson, *Glossary of Reliability and Maintenance Terms*, Houston, TX: Gulf Publishing Company, 1997.

[13] Terry, G. R., *Principles of Management*, Chicago, IL: Irwin, 1972.

[14] Dessler, G., *Management Fundamentals: Modern Principles and Practices*, Reston, VA: Reston Publishing Company, 1982.

[15] Thamhain, H. J., "Developing Engineering Program Management Skills," in *Handbook on Management of R and D and Engineering*, D. F. Kocaoglu (ed.), New York: John Wiley and Sons, 1992, pp. 400–421.

[16] Thamhain, H. J., "Managing Technology: The People Factor," *Technical and Skill Training*, August/September 1990, pp. 80–84.

2

Organizing and the Human Element

2.1 Introduction

Industrial organizations have been increasing in size, number, and the size of their operations from the days of the Industrial Revolution. Today's organizations have to operate in more demanding and competitive environments than the environments of the past. For example, 50 years ago companies could survive with only one or perhaps two product lines, but in today's environments their survival depends on factors such as multiple product lines (i.e., diversification) and vigorous integration of technology into the existing setups [1]. Factors such as these have been instrumental in organizing the operations of industrial organizations in such a manner that they operate most efficiently and at the minimum cost. The term *organizing* may simply be described as the process of developing a structure for the organization or company to help its manpower to operate systematically to satisfy the organizational goals in an effective manner [2].

The human element plays a pivotal part in the success or failure of an engineering company. For example, a company may have an excellent engineering product, but poor human management within the organization may make it an unprofitable product. Human resources are probably the most important resource of an organization because they are a major element of cost in most enterprises (e.g., in the petroleum and chemical industries, the labor cost varies between 25% and 30% of the total operating cost) and they influence an organization's productivity [3].

This chapter presents important aspects of the organizing and human element considered useful for engineering management professionals.

2.2 The Components of Organizing and Guidelines for Planning an Organization

The basic goal of organizing is to divide the work into reasonably manageable units. Irrespective of the permanency of such units, organizing involves many elementary components that require careful consideration during the design of any organizational system. Some of these components are as follows [1]:

- *Definition of work.* This calls for management to have a good understanding of the type, scope, and magnitude of work to be carried out in the future.

- *Division of work.* This is the process of subdividing the work under consideration into manageable units.

- *Division of labor and specialization.* This is concerned with allocating the subdivided work to groups and individuals specialized in carrying out the task/function.

- *Unity of command and direction.* One of the fundamentals of organization calls for an individual to receive orders from one superior authority only [4]. However, in modern organizational systems such as the matrix, dual accountability and power sharing could be unavoidable. Under such circumstances, each employee must be very clear concerning for which specific items he or she is responsible and to whom. The unity of direction calls for clearly directing each set of activities toward specified goals, unified by one accepted plan and one leader. In a modern organization, task direction and leadership channels are rather intricate, thus they must be carefully considered.

- *Responsibility and authority.* Each organizational unit and each individual must have clearly delineated authority defining the scope and type of work to be executed and how to work in harmony, which includes the authority to pass orders and use resources.

- *Chain of command and span of control.* A chain of command may be described as an unbroken line of authority that links all concerned individuals to successively higher authority levels. In establishing an organizational structure the aim is to provide a clear chain of command for each and every command axis of the organization. The span of control is the number of people reporting directly to a supervisor. Usually various factors are considered in establishing span of control at each organizational level.

- *Centralization.* This is concerned with the concentration of authority at the top level of an enterprise or any of its elements. Usually, the degree of centralization varies from one organization to another.

Past experience indicates that the following guidelines are quite useful in planning an organization [5]:

- Ensure that the individual delegating responsibility is accountable.
- Keep a position's functional and managing duties separate.
- Ensure that an individual receives orders from only one authority.
- Keep the management levels to a minimum.
- Keep the delegation of authority and responsibility close to the point of action as much as possible.
- Keep the line of responsibility and authority from the top to the bottom of the organization as clear as possible.
- Design each organizational element such that it satisfies its own objectives effectively in addition to that of its parent enterprise.

2.3 Organizational Charts and Basic Relationships in Organizational Structures

In an organization, there are probably only a handful of documents that get treated with more respect and emotion and carry more authority than organizational charts. Reasons for having organizational charts include the assignment of responsibilities to individuals, the outlining of basic authority, the identification of weak or strong control, and the outlining of basic relationships. In addition, organizational charts improve communication channels, simplify management functions, provide a sense of security, serve as a basis for directives, serve as reference documents, and serve as a framework for budgeting and scheduling [6].

Some of the limitations of the organizational charts include [1]:

- An inability to show work flow;
- An inability to show informal organization;

- An inability to display factors such as dual accountability, resource sharing, and power sharing;

- Inflexibility;

- A state of being too static;

- An inability to show how the work is being performed and integrated;

- An overemphasis on status and position.

There are three basic relationships in the organizational structures of many companies as shown in Figure 2.1: line, group, and staff. In the case of line, authority flows from superior A to subordinate B, and then through B to subordinate C. This way a line is formed from the top management level to the lowest organizational level. That is why it is known as the *line authority*. The group is also known as *multiple reporting* and in this case a group of individuals report to a single superior. Usually, more people report to a superior at the bottom organizational level than at the top level. In the case of the staff relationship, a group of individuals perform the role of helping the line management carry out the organizational goals in an effective manner. The staff relationship is not something new—in ancient times, Alexander the Great's (355–323 B.C.) armies practiced this concept [2]. Usually, the staff function or relationship exists more at the top end of the organizational ladder than at the bottom end.

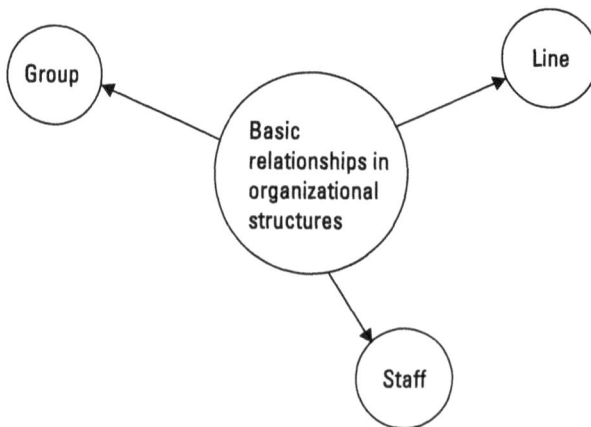

Figure 2.1 Three basic relationships in organizational structures.

2.4 Centralization and Decentralization of Organizations, Span of Control, and Delegation

Centralization and decentralization are important issues in establishing the organizational structures of engineering companies. In the case of centralization, the upper management has the authority to make the important or critical decisions. By contrast, in the case of decentralization, the authority is distributed among lower levels of the organization. Nonetheless, for an effective performance, a proper balance between the two is necessary.

Past experience indicates that an organization leans more toward centralization if the following conditions exist [1]:

- It is a capital-intensive operation.
- There are clear channels of supply and distribution.
- There is a need for close coordination of activities.
- There is a need for confidentiality of top-level decisions.
- There is a need for top-level strategy and direction.
- Available resources can be pooled and shared.
- There is a need for advanced technology and skill developments.

Some of the benefits of centralization are the elimination of work duplication, a reduction in the number of undesirable decisions by less experienced executives, an improvement in control over specialized functions, closer communication between the important decision makers (thus simplifying the coordination of their efforts), the causing of centralized staff expertise to be more efficient and simpler to use, a reduction in the need for experienced subordinate managers, and a reduction in staffing costs because fewer competent managers can handle the jobs and the needs of the staff [2, 5].

Under the conditions listed next, an organization that leans toward decentralization has the following characteristics [1]:

- It is a labor-intensive operation.
- There is noncritical resource sharing.
- Separable operations and profit accountability exist.
- ' It is a large organization in diversified businesses.

- There is a need for quick reaction time in relation to the external business environment.
- There is a need from operating-level management for entrepreneurial drive, initiative, and commitment.
- It is a project- or multiproduct-oriented operation.

Nonetheless, as per [7], the factors such as those listed next shape the degree of decentralization of authority:

- The business's dynamic nature;
- Management's thinking philosophy;
- Company size;
- Company history;
- Desire for independence by individuals and groups;
- Availability of properly trained managers to make decentralized decisions;
- Cost of decisions;
- Influences from external forces (e.g., government controls, income tax policies, and unions);
- Availability of control techniques for monitoring decisions made at the lower management levels.

Some of the benefits of decentralization are useful for making the overall organization stronger by facilitating the personal development of individuals, which is useful to stimulate initiative and identification with the firm, useful for creating better feelings of satisfaction among competent individuals, and useful for ensuring that the decisions are made by management individuals with the best experience of local conditions [2, 5].

Span of control may simply be described as the number of individuals directly reporting to a manager. Although it is impossible to calculate the actual number of individuals in an ideal span of control, over the years various experts have proposed various numbers of people to be supervised by a superior. For example, an ideal number of individuals to be supervised by a superior at the upper levels of management is between four and eight and at the lower levels between eight and fifteen. However, a survey of 100 large American firms conducted by the American Management Association revealed that executives reporting to company presidents varied from one to 24, and there were only 26 presidents supervising six or less people [7].

Factors such as those listed next help define the limits of individuals which can be supervised by a superior [1, 7].

- The superior's capacity to comprehend quickly;
- The complexity of the supervised task or function;
- The physical proximity of the subordinate's functions;
- The requirement for personal involvement by superior;
- The team or group characteristics of subordinate manpower;
- The overhead function's responsibilities, such as staffing, organizing, planning, and reporting;
- The integration needs among subfunctions;
- The similarity of the supervised task or function;
- The amount of personnel administration versus task leadership only.

After extensive research, the model Lockheed Missile and Space Company developed a span of control mode that takes into consideration the following factors [7–9]:

- The degree of coordination required;
- The organizational help available to superiors;
- The locations of individuals reporting to a superior;
- The degree of direction and control required by subordinate individuals;
- The type of department or unit management;
- The nature of work performed;
- The importance of planning and functions of superiors or organizational units, their complexity and time requirements;
- The similarity of functions carried out by subordinate individuals.

Each of the above variables is assigned a weight in determining the optimum span of control.

In the past, some attempts have been made to develop mathematical span of control models. For example, [10] defines the total number of leaders in a company as follows:

$$TL = \frac{WF(M^n - 1)}{M^n(M - 1)} \qquad (2.1)$$

where

TL	is the total number of leaders in a company.
n	is the total number of hierarchy levels in a company.
WF	is the company work force excluding supervisory personnel.
M	is the number of persons to be supervised by a superior or leader.

Since a company has only one president, we write

$$\frac{WF}{M^n} = 1 \qquad (2.2)$$

Substituting (2.2) into (2.1) yields

$$TL = \frac{(M^n - 1)}{(M - 1)} \qquad (2.3)$$

Using (2.2), we get

$$n = \frac{\log WF}{\log M} \qquad (2.4)$$

and

$$M = (WF)^{\frac{1}{n}} \qquad (2.5)$$

This mathematical model is subject to observations such as the organizational levels decrease with the increase in span of control, uniform span of control at all hierarchy levels, and n is inversely proportional to log M because of constant WF. All in all, mathematical models such as this could be useful in making decisions concerning the span of control.

Delegation is concerned with assigning responsibility and authority to subordinate individuals by a superior, and it is absolutely necessary because superiors work through others. The task of delegation is not easy, and it suffers from various managerial and subordinate obstacles. The managerial obstacles include poor confidence in subordinates, poor directing capability, overwariness of risk in tasks performed by subordinates, false beliefs that the superiors can do a better job than the subordinates, and poor control mechanisms to make management aware of impending difficulty [11].

By contrast, some of the subordinates' obstacles to delegation are poor self-confidence, poor incentives for additional responsibilities, heavy current workloads, poor facilities for accomplishing the job properly, fear of mistakes, and reluctance to sort out the problem themselves because they find it easier to obtain the help of superiors [11].

Over the years, various people have presented interesting approaches for delegating authority. Some of the useful guidelines for managers in delegating authority are as follows [8]:

- Choose only those individuals whose qualifications and experiences match the tasks in question.

- Divide the job under consideration into various separate tasks.

- Aim for the total understanding of subordinate individuals with respect to factors such as orders being delegated, his or her authority, the subsequent review procedure to be used, and subsequent reward (if any).

- Monitor progress periodically.

- Provide sufficient authority to subordinates.

- Describe the responsibility standards for each task to be performed.

2.5 Methods of Organization

Many different methods are used to develop organizational structures. The application of these methods depends on factors such as location, company policy, skills of manpower, and product. This section presents five methods of organization [2, 5–7].

2.5.1 Organization by Product

Some companies use this approach to develop their organizational structure. The approach calls for dividing the company into divisions and then assigning each division the responsibility for one particular product. Figure 2.2 shows a simplified chart of an organization by product.

Some of the benefits of this approach are as follows:

- It develops better coordination of functional activities.

- It places better attention on the product.

- It develops teamwork more easily.

- It places profit responsibility at the divisional level.

- It serves as a measurable training ground for general managers.

By contrast, the major disadvantages of this method are an increase in need for general managers, the limitation of contact among individuals of the same specialty, and the difficulty in maintaining cost effectiveness of central services.

2.5.2 Organization by Function

This is a widely used method in organizing company activities, and it calls for dividing work according to discipline or subject in addition to performing all similar work under one unit. This method is often favored by large research

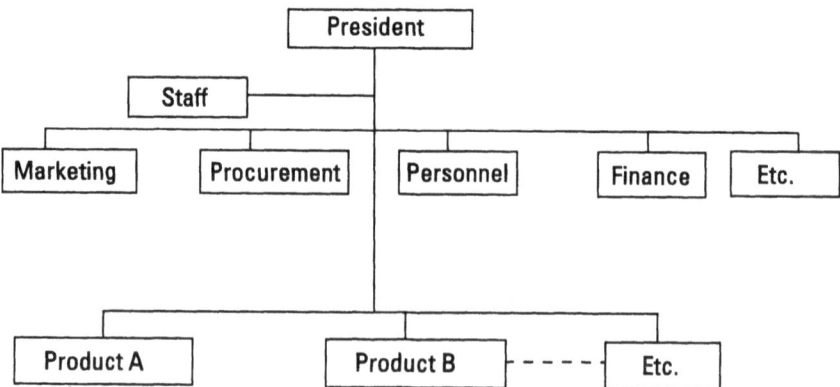

Figure 2.2 Organization by product.

groups and enterprises as well as organizations with long-term projects [6, 9]. Figure 2.3 presents a simplified chart of a functional organization.

Some of the advantages of this method are as follows:

- More uniform products;
- Consistent policy;
- Even distribution of work;
- Elimination of duplicate facilities;
- Room for technical specialization;
- Group homogeneity, thus it is easy to supervise a group.

The method also has several drawbacks, including slow work flow, difficulty when shifting personnel, and poor effectiveness in cross-discipline developmental work.

2.5.3 Organization by Project

In this case, appropriate individuals are grouped together to carry out a complex project within prescribed limits. Thus, the project organization may simply be described as a nonpermanent structure formed to meet a specific objective. More specifically, after the completion of the project the individuals associated with the project are either sent back to their permanent department or transferred to a new project.

Past experience indicates that small and medium-sized enterprises and companies with various short-term jobs favor the organization by project approach [6]. A simplified organizational structure of a project-based

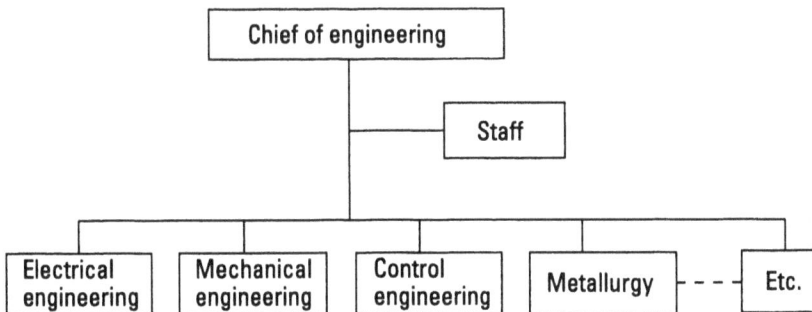

Figure 2.3 A functional organization.

organization is shown in Figure 2.4. Some of the benefits of the organization by project approach are as follows [1, 5, 6]:

- It focuses attention on a single project.
- It improves efficiency of work flow.
- It is a useful framework for team effort.
- It allows improved coordination of large projects.
- It is a useful tool for specialization by product.
- It is useful to make more specific the accountability and responsibility of the project.

By contrast, the disadvantages of this approach include less uniform products, duplication of work facilities, and inconsistent policy [9].

2.5.4   Organization by Territory

This approach is used in a situation where a company is physically dispersed [7, 9]. Under this approach, all activities in a given territory are grouped and then a manager is appointed to head the group. There are various reasons for having "departmentation" by territory, including poor communication facilities, a need to take prompt action, and the encouragement of local participation in decision making. Figure 2.5 presents a simplified diagram of organization by territory.

Some of the advantages of organization by territory are the emphasis placed on local conditions, usefulness in upgrading regional coordination, and the advantage taken of local conditions' economy. By contrast, the method's two main disadvantages are the difficulty in maintaining economical central services and the increased need for general managers.

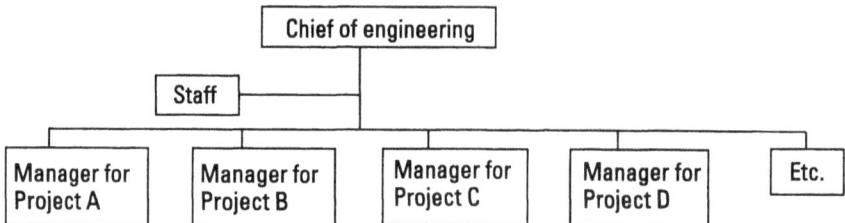

Figure 2.4 A simplified chart of a project-based organization.

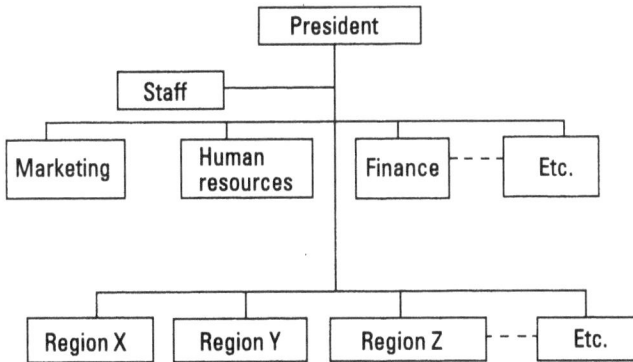

Figure 2.5 A simplified chart of organization by territory.

2.5.5 Matrix Organization

Broadly speaking, this is the result of combining the functional and project methods together. This way most advantages of both the approaches are achieved. The matrix organization method was first practiced by medium-sized aerospace companies in the 1950s because these organizations were not large enough to take full advantage of the organization by project approach. In the matrix organization, the project manpower is loaned to the project manager, and it reports to both the project manager and the chiefs of the functional or "home" departments. More specifically, the functional departments are the ones to which this manpower is permanently assigned.

Some of the benefits of the matrix organization are efficient decision making, which provides a better control and increases the role of middle management [12]. Similarly, two of its drawbacks are the possibility of producing surplus skills in the event of smaller projects and the application of such skills to projects that do not require it.

2.6 Functions of an Engineering Department and Guidelines for Organizing a New Engineering Department

A typical engineering department performs various types of functions, and they may be grouped under five distinct categories as shown in Figure 2.6 [6].

The administrative category includes functions such as those listed here:

- Formulating polices and planning;
- Hiring, firing, and promotion;

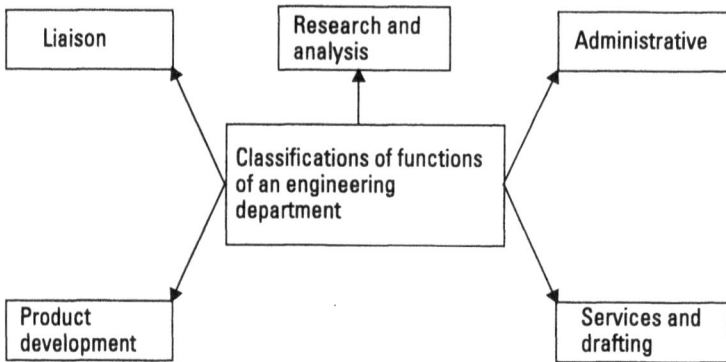

Figure 2.6 Categories of functions of a typical engineering department.

- Budgeting;
- Pricing;
- Assigning duties and controlling work;
- Procurement;
- Progress reporting;
- Training;
- Salary review.

The following functions belong to the research and analysis category:

- Searching for new ideas;
- Performing patent and literature searches;
- Testing and evaluating;
- Finding solutions to basic problems;
- Providing analytical services;
- Performing calculation and computing.

Some of the functions belonging to the product development classification are as follows:

- Preparing and analyzing proposals;
- Redesigning;

- Searching for new products;
- Standardizing parts and materials;
- Designing new products.

The functions such as those listed next belong to the services and drafting category:

- Releasing drawings and prints;
- Drafting;
- Constructing models;
- Publications.

Some examples of the functions belonging to the liaison category are as follows:

- Liaising with customers;
- Liaising with other departments;
- Liaising with outside agencies.

Over the years professionals working in the engineering management field have developed various guidelines for organizing a new engineering department. Some of these guidelines are as follows [1]:

- Develop the overall mission for the department by including factors such as technical objectives, global business objectives, timing, and responsibilities.
- Define the type of tasks to be carried out by the department.
- Define the functional capabilities that the new department should include versus the ones that can be contracted out to other departments or organizations. Ensure that the senior management agrees with the organizational scope prior to taking the next step.
- Define with care the main functional units of the department, along with its reporting structure.
- Define factors such as the specific reporting relations, controls, and responsibilities that will make the department operational.

- Delineate and summarize in the organizational plan in a step-by-step fashion the specific staffing levels and facility and equipment requirements.

- Develop job descriptions for all individuals who will report directly to the department manager.

- Advertise positions in the department to be filled, within and outside the company.

2.7 Characteristics and Needs of an Engineer, Routes for an Engineer to Obtain Management Positions, and Transition of an Engineer to a Managerial Position

An engineer possesses certain characteristics and some of the important ones include liking new and different things, being independent, being recognized, exercising technical knowledge and skills, building relationships with other engineering professionals, and directly attacking problems [13].

The many needs of an engineer include job security, acceptable work facilities, economic advancement, opportunities for self-development, work variety, appropriate technical assistance, stimulating and challenging work, adequate supporting staff, independence to attack a work problem, participation in decisions that will affect him or her, proper work assignment, competent bosses, employment with a reputable company, opportunity for his or her ideas to be practiced, having clearly defined responsibility and authority within an organization, and proper recognition for his or her efforts from the management [5, 9, 13]. There are many different routes for an engineer to obtain management positions as shown in Figure 2.7 [5].

The promotion is concerned with changing the current job after obtaining necessary experience, say, 2 or 3 years at one place, in the early stage of an engineer's career because in many organizations the chances of getting a promotion are rather low. Reasonable good service with a firm and leadership in professional activities is another route for a management position, although it is lengthy. The leadership in professional activities is obtained through actions such as writing articles or books and securing positions in professional societies. This allows an engineer a "high visibility" with the company management. A high degree of technical competence and ability to organize is one of the best routes to obtain management positions, particularly for persons doing extremely well in their professional specialties. The sponsor-protégé arrangement is another route for managerial positions. In this case, a manager finds dependable and likeable people and when such a

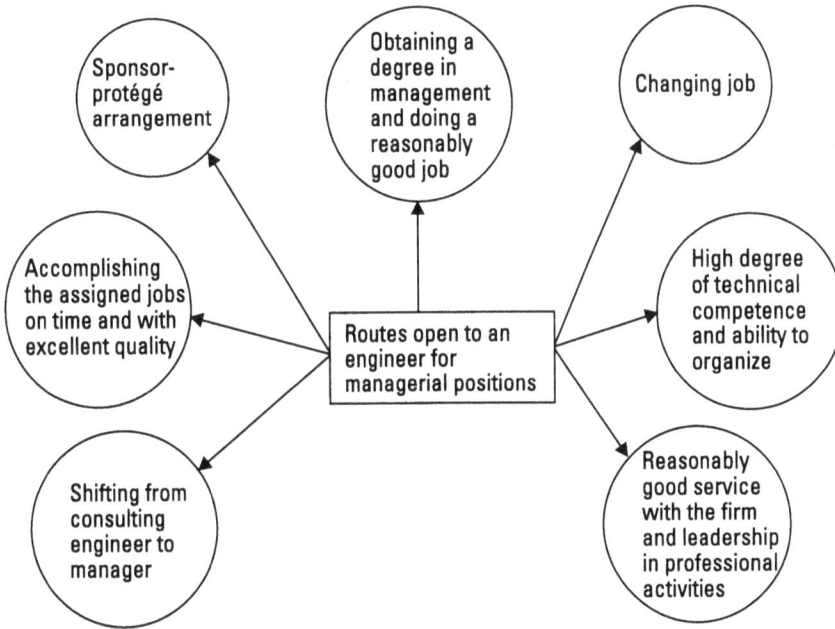

Figure 2.7 Routes for an engineer to obtain managerial positions.

manager gets a promotion, he or she tries to get promotions for his or her closest people, so that they can work for him or her in the new environment.

Accomplishing the assigned jobs on time and with excellent quality is another way of obtaining managerial positions. Obtaining an advanced degree in management and doing a reasonably good job in the field has become a fairly popular route for obtaining management positions. For example, a survey of technical professionals revealed that engineers and scientists who obtained a master's degree in management within 1 to 3 years of their college graduation earn around 17% higher than the ones with a master's degree in engineering [5, 9]. Engineers who develop their specialties by becoming consulting engineers can also move to managerial ranks. Usually, this route works quite well for consulting engineers who do not become too specialized.

Whenever an engineer makes a switch from a technical position to a management position, he or she finds certain changes in areas such as responsibilities and personnel habits. Therefore, such an individual must be prepared for these changes; otherwise, he or she may find it difficult to perform his or her job effectively. Some of these changes are listed next [6].

- *Dealing with generalities.* This simply means the engineering professional has to work with generalities such as supervision, delegation, sales, and negotiation instead of specifics such as force, pressure, weight, and length in engineering.

- *Reading habits.* The engineer now not only has to read technical magazines but also the management ones. However, due to lack of available time, the individual will look for results rather than working to understand each and every step of an approach or process described in these documents.

- *Human relations.* Now the engineer not only has to look after his or her own interests but also those of the people he or she supervises. He or she will be concerned with maximizing productivity from people with minimum friction and effort.

- *Delivering speeches.* Because management solves many of its problems in meetings, a supervising engineer or manager is expected to propose, explain, and defend his or her ideas in such gatherings. From time to time, a manager may also make speeches to public audiences. All in all, a person in a managerial position has to learn to think and talk on his or her feet in order to advance further in his or her career.

- *Delegating.* As no manager can carry out all the tasks required from his or her department or group, such tasks must be delegated to others within the department. Thus, a person new to a management position has to learn the secrets of delegating tasks pleasantly and effectively.

- *Changing ways of thinking.* An engineer's thinking is usually confined to only one technical project, but in managing his or her thinking will be broader. For example, he or she must now be concerned with his or her department's performance and profit or loss to the overall company. Moreover, he or she will think in terms of selling to others, will think ahead about problems, will think of many alternatives, and will think while listening to others.

- *Training people.* Training others will be important to the newly appointed engineer-manager due to rapid changes in current modern technology. The people under his or her direction must be able to perform their jobs effectively.

2.8 An Engineering Manager's Qualities and Activities

Over the years, many professionals working in the management field have studied the qualities of engineering managers. After a careful analysis, they have developed a list of typical qualities of a good engineering manager as presented in Table 2.1 [6].

Usually, a manager performs the following basic activities [6, 9, 13]:

- Planning;
- Organizing;
- Staffing;
- Communicating;
- Personnel development;

Table 2.1
Typical Qualities of a Good Engineering Manager

Number	Quality
1	Tolerance
2	Flexibility
3	Fairness
4	Empathy
5	Ability to reason
6	Good emotional control
7	Good humor
8	Self-confidence
9	Good listening ability
10	Quickness to praise and criticize
11	Tact
12	Technical competence
13	Quickness to see good in others
14	Ability to recognize different points of view
15	Ability to self-appraise
16	Freedom from suspicion and prejudice
17	Good communication skills

- Counseling;
- Training;
- Standards.

Each of these activities is described in [6].

2.9 Motivating Others and Analyzing Team Characteristics

One way to increase the productivity of a department is to motivate its manpower effectively. Some past studies indicate that self-motivated individuals are two to ten times more productive than their non-self-motivated counterparts. Although many different motivational tools have been cited in published literature, as per [6], the following six motivators have proven to be very useful for managers to motivate people in the past:

- Involve employees in the goal-setting act.
- Set performance measuring standards high but within achievable limits of employees.
- Ensure that the employees clearly understand goals and policies.
- Challenge self-defeating attitudes of employees.
- Consult employees on matters that are usually above their level from time to time.
- Uncover motivational tools that work best for an individual employee.

Nonetheless, as per [3], some of the high motivators for engineers are making a direct attack on problems, performing a job successfully, being independent, associating with competent coworkers, doing new and different things, utilizing professional information to solve problems, using technical knowledge and skills in making contributions to the advancement of science, and being recognized for accomplishments by peers and colleagues.

Similarly, the important motivators for managers are having a challenging job, threat of competition, the job as a status symbol, urge for leadership, monetary gain, fear of losing job, and making mistakes [7].

Table 2.2 presents a list of useful questions for engineering managers to use to analyze team characteristics [13].

Table 2.2
Useful Questions for Analyzing Team Characteristics

Number	Question
1	What things do the team members do well?
2	What are the major drawbacks of the team?
3	How well do the team members organize their work?
4	How does the team interact with or relate to other groups in work or social situations?
5	Is there any warmth and friendliness among team members?
6	What interpersonal relationships exist among team members?
7	Do team members clearly understand such factors as their responsibilities, plans, and roles?
8	What is the attitude of the team members toward dominant requirements, and what are they doing to meet them? Is there anything that turns them off?
9	What is the team's past experience?
10	Are there any initiators among team members who try to get other members to join in doing things?
11	Are there any distinct motivational patterns of the team members?
12	What is the main goal of the team, and is it clearly understood by the team?
13	What personal goals and values are important to team members?
14	What is really going on for the team?

2.10 Committees and Staff Meetings

In government and private sectors, committees are widely used to make various types of decisions. A committee is made up of individuals to whom, as a team, specific matter is committed [7].

There are various reasons for having committees: to gain group judgment, to reduce the power of a single individual, to represent different concerned groups, to share and transmit information, to coordinate activities among various groups of an organization, to encourage motivation through participation, to delay decisions on a problem, and so on [7, 9].

Some of the major functions of a committee are as follows [9]:

- Evaluating policies;
- Reviewing performance;

- Producing alternatives;

- Recommending solutions;

- Implementing solutions.

Just like anything else, committees too have various disadvantages, including being subject to minority tyranny, dividing responsibility, being costly in monetary terms, and taking considerable time to reach group decisions. Often various voices against committees are raised, but a survey published in the *Harvard Business Review* found that only 8% of the respondents would eliminate committees if they were ever given power to do so [14].

Staff meetings consume a significant amount of managers' time, and many important decisions are made in these meetings. A competent manager attends these meetings not only to discuss and find solutions to work problems, but also to discover and develop the management potential of his or her subordinates so that at a time of need, the organization will find little difficulty in looking for people with management talent. From the manager's perspective, some of the benefits associated with the staff meetings are as follows [6]:

- They are useful in identifying the fast thinkers.

- The manager will not overlook the potentially gifted executives.

- The manager can monitor his or her entire staff in action.

- Staff meetings are useful testing grounds for discovering the management capabilities of an individual. For example, a person's performance during the staff meetings will give a fair idea about that person's management potential.

- Staff meetings are useful for monitoring the attitudes of staff members.

The staff meetings can only be useful if they are held effectively. Some useful guidelines for holding effective staff meetings are as follows [6, 8, 9]:

- Control the meeting by not allowing irrelevant topics to dominate, leaving the meeting's objective unaccomplished.

- Stimulate the interest of staff members in meetings so that they do not consider them a waste of time.

- Do not permit any disruptions during meetings.

- Encourage the participation of all persons present at the meeting through actions such as circulating the meeting agenda well in advance.

- At the start of the meeting, examine the actions taken on decisions of the previous meeting.

- Promote the use of visual aids, particularly when presenting facts and figures.

- Tactfully, assign seats to participants, particularly to a potential troublemaker, whenever the manager (i.e., the chairperson) has an opportunity. As per the behavioral scientists [8], the potential troublemaker should be assigned seat "T" as shown in Figure 2.8. As per the statistics, the manager chairing the meeting is usually a right-eyed dominant person, thus he or she can tactfully ignore the troublemaker seated at position "T." However, when a person wishes to challenge the chairperson, he or she should sit on seat "C" as shown in Figure 2.8. This allows the challenger to confront the person chairing the meeting with eye-to-eye contact, and it could be quite disconcerting to the chairperson if the challenger keeps his or her hand up to participate in the discussion. All in all, a manager or others chairing a meeting should avoid such a seating arrangement.

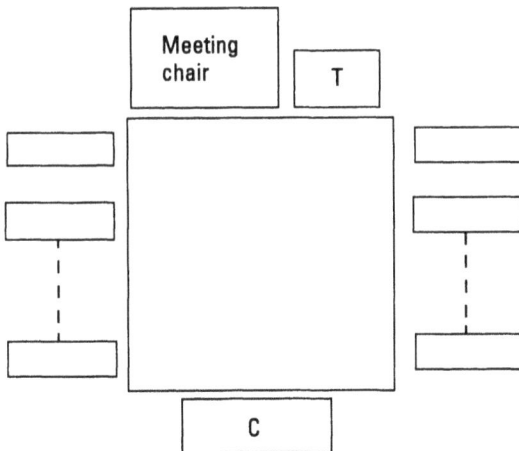

Figure 2.8 Seating layout for a meeting.

2.11 Displacing Managers and an Organization Size-Efficiency Model

From time to time, management finds itself in a situation of displacing managers due to various factors including poor performance. Displacing a manager from his or her position is not simple or straightforward—it requires careful execution. Otherwise, the company image in the eyes of its employees, customers, and the public could be severely damaged. Past experience indicates that senior management practices various ways of displacing managers. Figure 2.9 presents five basic ways [15]. Each of these ways is discussed separately next.

Center

In this case, the concerned manager's title and office are retained but his or her group is removed or shifted to another department. Usually, this move is practiced in a situation where it is necessary to conceal from the public the removal of the manager. However, this move has several drawbacks: the concerned individual's salary may be maintained at the same level and the morale of the shifted group members may decrease along with a decrease in productivity.

East

In this case, the affected manager is moved laterally to a position such as a special or staff assistant with the same salary. If the conditions change, the involved individual may return to his or her original position. Some of the benefits of this move are as follows:

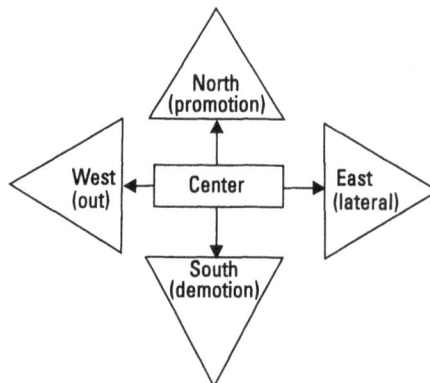

Figure 2.9 Five basic ways to displace a manager.

- Often, the affected individual is receptive to the move.
- The move is relatively easy to execute.
- Chances for publicity are minimal.
- The morale and loyalty of the involved manager to the firm may not be affected, if the lateral position will be of real value to the individual in question.

North

This is probably the easiest move to remove a manager from his or her current position, and it involves giving such an individual a promotion. Usually, this move is practiced in situations where the manager has become powerful by having strong links with the public, customers, and so on. There is a good chance that the affected manager may not even become aware of the hidden motive and may take it as a genuine promotion.

Two important advantages of the move are that the morale of the individual and the group reporting to him or her remains intact and that it is easy to get the approval of the involved person because of a more prestigious title and probably higher financial benefits.

By contrast, the following two principal drawbacks are associated with this move:

- It is costly because the salary is higher or the same.
- The affected manager may interfere with the functioning of the old group because of the group's loyalty toward him or her.

South

In this case, the concerned individual is demoted from his or her current position. Usually, this move is practiced with an individual who will take it well or is old or sick. Past experience indicates that young individuals do not usually take such humiliation easily and remain loyal to the company. Two principal advantages of this move are probable savings in the payment of salary and the removal of the individual from his or her position. Similarly, the move's disadvantages include probable reduction in an individual's productivity, a lesser degree of loyalty toward the organization, and a possible poor effect on the morale of other members of the company.

West

In this case, the manager is asked to leave the company. Normally, this move is practiced in situations when the other moves fail. From time to time, the moves such as those discussed earlier are practiced first with the hope that the concerned individual leaves the organization at his or her will. Also, sometimes this move is softened with an offer of early retirement or disability leave (if applicable).

Some of the disadvantages of the move are increased chances of adverse publicity, the possibility of a poor effect on the morale of certain members of the organization, and the possibility of an adverse effect on customers and investors because they may interpret this as the existence of unstable forces in the company.

2.11.1 An Organization Size-Efficiency Model

This model directly or indirectly studies both the organizational and human element aspects. More specifically, the model is basically concerned with determining the relationship between organization size and employees' output. The model is based on the assumption that the organization is involved in paper study oriented tasks such as doing research. More specifically, a professional employee reads other people's works or reports during work hours and writes his or her reports. The following expression was developed to determine the number of publications or reports produced by an organization or department annually [16]:

$$N = \frac{WDM}{\left(AT_w + AT_r M\theta\right)} \tag{2.6}$$

where

N is the total number of publications or reports produced by an organization or a department per year.

AT_r is the average time taken to read one report by a professional employee, expressed in days.

AT_w is the average time taken to write one report by a professional employee, expressed in days. This time incorporates the time spent on items such as investigation, analysis, writing, and review.

θ is the fraction of all reports or publications received by a professional employee, more specifically, the documents the employee is expected to read.

WD is the assumed number of working days in a year. Thus, in this model, $WD = 240$ days.

M is the total number of professional employees in the organization or department.

By inserting the value of WD into (2.6), we get

$$N = \frac{240M}{(AT_w + AT_r\,M\theta)} \tag{2.7}$$

As M becomes very large, (2.7) yields

$$N' = \frac{240}{AT_r\,\theta} \tag{2.8}$$

where N' is the upper limit of N.

The efficiency of the organization or department is expressed by

$$\gamma = \frac{N}{N_O} = \left[1 + (\theta MAT_r)\,/\,AT_w\right]^{-1} \tag{2.9}$$

where N_O is the total number of reports that can be produced if no time was spent in reading any of the reports written by others.

From (2.9), it can be easily noted that as M becomes large, the value of γ approaches zero, γ being inversely proportional to M.

Example 2.1

Assume that a research organization employs 400 professional employees, and each employee works for 7 hours a day throughout 240 workdays annually. Each employee is involved in a research activity in which he or she reads other people's works in order to write his or her own reports. Past experience indicates that on the average, the employee spends 5 days reading a report prepared by others and 30 days writing his or her report. The time spent for writing includes tasks such as investigation, analysis, and actual writing. If each professional employee reads only 25% of the total reports received annually, calculate the following:

- The total number of reports expected to be produced by the organization per year.
- The total number of reports expected to be produced by the organization annually if the total number of employees in the organization increases to 800 and 1,600.

For M = 400, 800, and 1,600, substituting the other given data into (2.7) yields the following results, respectively:

$$N = \frac{240\,(400)}{30 + 5\,(400)\,(0.25)}$$
$$= 181.13 \text{ reports/year}$$

$$N = \frac{240\,(800)}{30 + 5\,(800)\,(0.25)}$$
$$= 186.41 \text{ reports/year}$$

and

$$N = \frac{240\,(1{,}600)}{30 + 5\,(1{,}600)\,(0.25)}$$
$$= 189.16 \text{ reports/year}$$

By substituting the specified data values into (2.8), we get

$$N' = \frac{240}{5\,(0.25)}$$
$$= 192 \text{ reports/year}$$

The above results clearly demonstrate that doubling the company manpower (i.e., 800 and 1,600) did not double the output. In fact, it went from 181.13 to 186.41 and then to 189.16 reports per year. The above results also show that the outputs are increasing towards the upper limit of 192 reports per year.

2.12 Problems

1. List and discuss the important components of organizing that require a careful consideration during the design of any organizational system.
2. What are the advantages of having organizational charts?
3. Discuss the factors under which an organization should be considered for decentralization.

4. What are the important factors that should be considered in determining the number of people to be supervised by an individual?

5. Discuss the following methods of organization:

- Organization by function;

- Organization by project.

6. What are the functions of a typical engineering department?

7. What are the characteristics and needs of an engineer?

8. What are the qualities of a good engineering manager?

9. What are the functions of a typical engineering manager?

10. Write down at least 10 useful questions for analyzing team characteristics.

11. Discuss at least five important guidelines for holding staff meetings effectively.

12. What are the important reasons for having committees?

References

[1] Thamhain, H. J., *Engineering Management*, New York: John Wiley and Sons, 1992.

[2] Mescon, M. H., M. Albert, and K. Khedouri, *Management: Individual and Organizational Effectiveness*, New York: Harper and Row Publishers, 1981.

[3] Glueck, W. F., *Personnel: A Diagnostic Approach*, Plano, TX: Business Publications, 1982.

[4] Fayol, H., *Industrial and General Management*, London: Pitman and Sons, 1949.

[5] Karger, D. W., and R. G. Murdick, *Managing: Engineering and Research*, New York: Industrial Press, 1969.

[6] Chironis, N. P. (ed.), *Management Guide for Engineers and Technical Administrators*, New York: McGraw-Hill, 1969.

[7] Koontz, H. C., C. O'Donnell, and H. Weihrich, *Management: A Book of Readings*, 5th ed., New York: McGraw-Hill, 1980.

[8] Kelly, J., *How Managers Manage*, Englewood Cliffs, NJ: Prentice Hall, 1980.

[9] Dhillon, B. S., *Engineering Management*, Lancaster, PA: Technomic Publishing Company, 1987.

[10] Meij, J. L., "Some Fundamental Principles of a General Theory of Management," *The Journal of Industrial Economics*, Vol. 1, October 1955, pp. 16–32.

[11] Newman, W. H., "Overcoming Obstacles to Effective Delegation," *Management Review,* January 1956, pp. 36–41.

[12] Pywell, H. E., "Engineering Management in a Multiple (Second- and Third-Level) Matrix Organization," *IEEE Trans. on Engineering Management,* Vol. 26, 1979, pp. 51–55.

[13] Amos, J. M., and B. R. Sarchet, *Management for Engineers,* Englewood Cliffs, NJ: Prentice Hall, 1981.

[14] Tillman, R., "Committees on Trials," *Harvard Business Review,* Vol. 38, No. 3, 1960, pp. 162–173.

[15] Bird, F. L., "How to Displace Executives," in *Management Guide for Engineers and Technical Administrators,* N. P. Chironis (ed.), New York: McGraw-Hill, 1969, pp. 63–64.

[16] Adler, F. P., "Relationships Between Organization Size and Efficiency," *Management Science,* Vol. 7, No. 1, 1960, pp. 80–84.

3

Tools for Making Effective Engineering and Technology Management Decisions

3.1 Introduction

Over the years professionals working in the management field have been trying to develop various types of tools for use in management activities. In particular, since World War II, there has been a considerable growth in the development of quantitative management techniques and methods. The new developments in computer technology and the complexities of modern management decision making problems have enhanced the importance of many of these quantitative approaches. Some examples of these methods are linear and nonlinear programming techniques.

Today's engineering and technology managers make use of various types of methods, including the ones developed for use in general management decision making. This chapter presents many approaches taken from published literature considered useful for making effective engineering and technology management decisions. These approaches include discounted cash flow analysis, depreciation analysis, decision trees, optimization techniques, learning curve analysis, fault tree analysis, and forecasting.

3.2 Discounted Cash Flow Analysis

In many engineering and technology management decisions, the time value of money plays an important role. This section presents some fundamental aspects of engineering economics useful to perform discounted cash flow analysis.

3.2.1 Compound Interest and Present Value

In the case of compound interest, at the end of each specified period, the earned interest is added to the original amount or principal. This new principal or amount acts as a principal for the next specified period and the process continues. Consequently, the interest is compounded into principal. Mathematically, at the end of the first period the amount is expressed by

$$A_1 = A + A_i$$
$$= A(1+i) \tag{3.1}$$

where

i	is the interest rate per period (usually, it is a year).
A	is the original amount or principal.
A_1	is the compound amount at the end of the first period or year.

At the end of the second period or year, the amount is

$$A_2 = A_1(1+i)$$
$$= A(1+i)^2 \tag{3.2}$$

where

A_1	is the principal amount for the second period or year.
A_2	is the compound amount at the end of the second period or year.

Similarly, at the end of the third period or year, the amount is expressed by

$$A_3 = A_2(1+i)$$
$$= A(1+i)^3 \tag{3.3}$$

where

A_3 is the compound amount at the end of the third period or year.

A_2 is the principal amount for the third period or year.

At the end of nth period (or year), the amount is generalized to the following form:

$$
\begin{aligned}
A_n &= A_{n-1}(1+i) \\
&= A(1+i)^n
\end{aligned}
\tag{3.4}
$$

where

A_n is the compound amount at the end of the nth period or year.

A_{n-1} is the principal amount for the nth period or year.

The total compound interest, I, earned after the nth period or year is expressed by

$$I = A_n - A \tag{3.5}$$

By rearranging (3.4), we get the following expression for the present value of a single payment:

$$A = A_n (I + i)^{-n} \tag{3.6}$$

In the above equation, A is the present value of A_n.

Example 3.1

Assume that a company has procured a robotic system and after 15 years of usage its value is estimated to be $80,000. The annual compound interest rate is estimated to be 5%. Calculate the present worth of the robotic system's estimated value of $80,000.

By inserting the given data values into (3.6), we get

$$A = (80{,}000)(1 + 0.05)^{-15}$$

Thus, the present worth of the robotic system's $80,000 value is $38,481.37.

3.2.2 Uniform Periodic Payment Amount and Present Value

In this case, it is assumed that at the end of each of the n equal periods or years, the depositor adds X amount of money. In turn, this amount is invested at the compound interest rate i. Thus, the total amount is expressed by [1, 2]

$$TA = X(1+i)^{n-1} + X(1+i)^{n-2} + - - - + X(1+i) + X \qquad (3.7)$$

where

TA	is the total amount of money.
$X(1+i)^{n-1}$	is the value of X after $(n-1)$ periods or years.
$X(1+i)^{n-2}$	is the value of X after $(n-2)$ periods or years.
$X(1+i)$	is the value of X after one period or year.

Equation (3.7) is a geometric series. To obtain its sum we multiply both sides of (3.7) by $(1+i)$:

$$TA(1+i) = X(1+i)^{n} + X(1+i)^{n-1} + - - - + X(1+i)^{2} + X(1+i) \quad (3.8)$$

By subtracting (3.7) from (3.8), we get

$$TA = \frac{X\left[(1+i)^{n} - 1\right]}{i} \qquad (3.9)$$

In a similar manner to writing (3.7), we write the following equation for the present value of uniform periodic payments:

$$PV_S = \frac{X}{(1+i)} + \frac{X}{(1+i)^{2}} + \frac{X}{(1+i)^{3}} + - - - + \frac{X}{(1+i)^{n}} \qquad (3.10)$$

where PV_S is the present value of uniform periodic payments.

The right side terms of (3.10) are the present values of X for 1, 2, 3, −, n years. Again (3.10) is a geometric series.

To obtain its sum we multiply both sides of (3.10) by $\left(\dfrac{1}{1+i}\right)$

$$\frac{PV_S}{(1+i)} = \frac{X}{(1+i)^{2}} + \frac{X}{(1+i)^{3}} + \frac{X}{(1+i)^{4}} + - - - + \frac{X}{(1+i)^{n+1}} \qquad (3.11)$$

Subtracting (3.10) from (3.11) yields

$$\frac{PV_S}{(1+i)} - PV_S = \frac{X}{(1+i)^{n+1}} - \frac{X}{(1+i)} \tag{3.12}$$

By rearranging (3.12), we get

$$PV_S = \frac{X\left[1-(1+i)^{-n}\right]}{i} \tag{3.13}$$

Example 3.2

Assume that an individual deposits $40,000 at the end of each year for the next 5 years. The annual compound interest rate is 7%. Calculate the total amount and the present value of the money deposited after the 5-year period.

By substituting the given data into (3.9), we get

$$TA = \frac{(40,000)\left[(1+0.07)^5 - 1\right]}{0.07}$$
$$= \$230,029.6$$

Using the specified data values in (3.13) yields

$$PV_S = \frac{(40,000)\left[1-(1+0.07)^{-5}\right]}{0.07}$$
$$= \$164,007.9$$

The total amount and present value will be $230,029.6 and $164,007.9, respectively.

3.3 Depreciation Analysis

The meaning of the term *depreciation* is decline in value. As engineering systems become older, they decline in value. Thus, depreciation analysis is basically concerned with considering the change in the value of items.

Over the years various methods for determining depreciation have been developed and this section presents three such methods [3].

3.3.1   Method I

This is often referred to as the *sum-of-the-years-digits* (SYD) method. This name is due to the calculation procedure used. The approach provides a larger depreciation charge in the early years of the item's useful life than in the later years. The annual depreciation charge is expressed by

$$ADC = \frac{RYIL\,(PC - SV)}{SYD \text{ for the total life}}$$
$$= \frac{(L - BY + 1)(PC - SV)}{\displaystyle\sum_{i=1}^{L} i} \tag{3.14}$$

where

ADC	is the annual depreciation charge.
RYIL	is the remaining years of item life.
PC	is the procurement cost of item.
L	is the item's useful life.
BY	is the year of item book value.
SV	is the salvage value.

The sum of the denominator of (3.14) is

$$\sum_{i=1}^{L} i = \frac{L(L+1)}{2} \tag{3.15}$$

Using (3.15) in (3.14) yields

$$ADC = \frac{2(L - BY + 1)(PC - SV)}{L(L+1)} \tag{3.16}$$

The item book value at the end of year *BY* is given by

$$BV_{BY} = \frac{2(PC - SV)\left[1 + 2 + \text{---} + (L - BY)\right]}{L(L+1)} + SV \tag{3.17}$$

Example 3.3

Assume that a company has procured a machine for $80,000 and its expected useful life is 15 years. The estimated salvage value of the machine is $5,000. Determine the annual depreciation charge during the fourth year of use with the aid of (3.16).

By inserting the specified data values into (3.16), we get

$$ADC = \frac{2(15 - 4 + 1)(80,000 - 5,000)}{15(15 + 1)}$$

$$= \$7,500$$

The annual depreciation charge during the fourth year of machine use will be $7,500.

3.3.2 Method II

This is known as the declining-balance depreciation method and it accelerates write-off item cost in early productive years and the years less close to the final life. The *depreciation rate* (DR) is expressed by

$$DR = 1 - \left(\frac{SV}{PC}\right)^{\frac{1}{L}} \tag{3.18}$$

This method calls for a positive value of *SV*. The item book value at the end of year *BY* is expressed by

$$BV_{BY}(BY) = PC(1 - DR)^{BY}$$
$$= (PC)\left[\frac{SV}{PC}\right]^{\frac{BY}{L}} \tag{3.19}$$

The annual depreciation charge, ADC_{BY}, at the end of year *BY* is expressed by

$$ADC_{BY} = BV_{BY}(BY - 1)\left[1 - \left(\frac{SV}{PC}\right)^{\frac{1}{L}}\right] \tag{3.20}$$

Example 3.4

Assume that the procurement cost, the expected useful life, and the salvage value of an engineering item are $50,000, 10 years, and $4,000, respectively. Determine the item book value at the end of 5 years by using (3.19).

By using the given data values in (3.19), we obtain

$$BV_{BY}(5) = 50,000\left[\frac{4,000}{50,000}\right]^{\frac{5}{10}}$$
$$= \$14,142.13$$

The item book value at the end of 5 years will be $14,142.13.

3.3.3   Method III

This is known as straight-line depreciation method and is probably the most widely used approach in depreciation studies. The method assumes constant annual depreciation during the item's useful life. Thus, the annual depreciation charge is expressed by

$$ADC = \frac{PC - SV}{L} \tag{3.21}$$

The item book value at the end of BY is

$$BV_{BY}(BY) = PC - \left[\frac{(PC - SV)BY}{L}\right] \tag{3.22}$$

Example 3.5

A company procured electrical equipment for $500,000 and its expected useful life is 10 years. At the end of 10 years, the equipment's salvage value is estimated to be $40,000. By using the straight-line depreciation approach, determine the electrical equipment's annual depreciation charge.

Inserting the given data into (3.21) yields

$$ADC = \frac{500,000 - 40,000}{10}$$
$$= \$46,000$$

Thus, the annual depreciation charge is $46,000.

3.4 Decision Trees

A decision tree may simply be described as a schematic diagram of a sequence of alternative decisions and the end conclusions of those decisions. Decision trees are used to handle sequential problems. Some of the advantages of using decision trees are as follows [4]:

- They provide an effective pictorial mechanism to represent the sequential decision process.
- They make the computations of expected value easier.
- They permit the consideration of the actions of more than one decision maker.

Two symbols used in the construction of decision trees are shown in Figure 3.1 [5].

The rectangle denotes a decision node from which one of many alternatives may be chosen. The circle represents a state of nature node out of which one state of nature will occur.

During the construction of a decision tree, one must assure that all alternatives and states of nature are in their proper and logical places in addition to including all possible alternatives and states of nature. The following steps are involved in analyzing problems with decision trees [5]:

- Define the problem under consideration.
- Structure or construct the decision tree.
- Assign probabilities to the states of nature.
- Determine payoffs for each and every feasible combination of alternatives and states of nature.

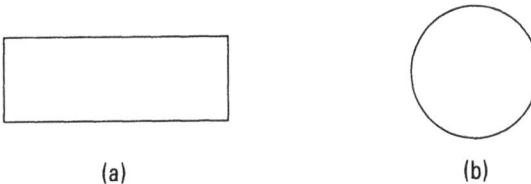

(a)   (b)

Figure 3.1 Decision tree symbols: (a) rectangle and (b) circle.

- Solve the problem under consideration through the computation of expected monetary values for each state of nature node. This is accomplished by working in a backward direction—by starting at the right side of the decision tree and working backward to decision nodes on the left side.

The following example demonstrates the decision tree concept [6–8]:

Example 3.6

An engineering manufacturer has the option of developing either product X or Y. There is about 70% probability that product X will generate a revenue of $2 million and about 30% chance that it will lose $500,000. If product X is successful, the manufacturer can also develop product A with 80% probability of earning $1.5 million and a 20% chance of losing $200,000.

By contrast, product Y has 90% probability of earning $2.5 million and a 10% chance of losing $1 million. If product Y is successful, the company has an opportunity to develop another product, B, with 95% probability of earning $1.5 million and a 5% chance of losing $800,000.

Product A and B will only be produced if products X and Y, respectively, are successful. Determine the product the manufacturer should consider for development.

Figure 3.2 presents a decision tree for Example 3.6. The acronyms FE, FL, and m in the figure stand for future earning, future loss, and million, respectively, and the numbers in parentheses denote probability. The diagram of Figure 3.2 shows six paths originating from the rectangle.

The paths 1 to 6 probabilities of occurrence are as follows:

- Path 1 probability of occurrence = (0.7) (0.8) = 0.56;
- Path 2 probability of occurrence = (0.7) (0.2) = 0.14;
- Path 3 probability of occurrence = (0.3) = 0.3;
- Path 4 probability of occurrence = (0.9) (0.95) = 0.855;
- Path 5 probability of occurrence = (0.9) (0.05) = 0.045;
- Path 6 probability of occurrence = (0.1) = 0.1.

Similarly, the paths 1 to 6 monetary outcomes are as follows:

- Path 1 monetary outcome = 2 + 1.5 = $3.5 million;

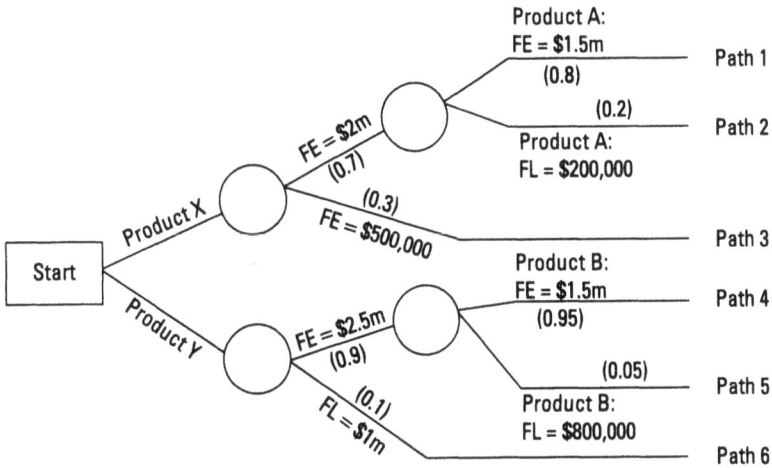

Figure 3.2 Decision tree for Example 3.6.

- Path 2 monetary outcome = 2–0.2 = $1.8 million;
- Path 3 monetary outcome = $500,000 (loss);
- Path 4 monetary outcome = 2.5 + 1.5 = $4 million;
- Path 5 monetary outcome = 2.5 – 0.8 = $1.7 million;
- Path 6 monetary outcome = $100,000 (loss).

The following expected values of paths 1 to 6 are obtained by multiplying the above corresponding path occurrence probabilities and monetary outcomes:

- Path 1 expected value = (0.56) (3.5) = $1.96 million;
- Path 2 expected value = (0.14) (1.8) = $252,000;
- Path 3 expected value = (0.3) (0.5) = $150,000 (loss);
- Path 4 expected value = (0.855) (4) = $3.42 million;
- Path 5 expected value = (0.045) (1.7) = $76,500;
- Path 6 expected value = (0.1) (0.1) = $10,000 (loss).

Using the above result, the *expected monetary value* (EMV) of the decision to develop product X is

$$EMV = 1.96 + 0.252 - 0.15$$
$$= \$2.062 \text{ million}$$

Similarly, the EMV to develop product Y is

$$EMV = 3.42 + 0.0765 - 0.01$$
$$= \$3.4865 \text{ million}$$

By examining these two results, it is concluded that the manufacturer will be better off by $1.4245 million if product Y is selected for development.

3.5 Optimization Techniques

There are many optimization techniques, and two of the best known methods are Lagrangian multiplier and linear programming. The knowledge of these two methods is considered useful to engineering management professionals. Both of these approaches are described next [5, 7].

3.5.1 Lagrangian Multiplier Method

This method is named after Joseph Lagrange, a French mathematician, who discovered it in the eighteenth century. The Lagrangian multiplier approach allows the optimization of functions subject to constraints without the elimination of any variables. The method is demonstrated for a two-variable function but in a similar manner, it can be extended for m variables. Thus, we define function $f(x_1, x_2)$ subject to the constraint function $g(x_1, x_2) = 0$. In this case, we write the Lagrange function as follows [9]:

$$L(x_1, x_2, \lambda) = f(x_1, x_2) + \lambda g(x_1, x_2) \qquad (3.23)$$

where

λ is the Lagrange multiplier.

$L(x_1, x_2, \lambda)$ is the Lagrange function.

When (3.23) is partially differentiated with respect to each of its arguments and then equated to zero, the necessary conditions for maximum or minimum are obtained. More specifically,

$$\frac{\delta L(x_1, x_2, \lambda)}{\delta x_1} = 0 \qquad (3.24)$$

$$\frac{\delta L(x_1, x_2, \lambda)}{\delta x_2} = 0 \qquad (3.25)$$

$$\frac{\delta L(x_1, x_2, \lambda)}{\delta \lambda} = 0 \qquad (3.26)$$

Expressions for x_1, x_2, and λ are obtained by solving (3.24)–(3.26).

Example 3.7

Determine the critical point for $f(x_1, x_2) = x_1^2 + x_2^2$ subject to constraint

$$g(x_1, x_2) = 3x_1 - x_2 - 6 = 0.$$

For this problem, we write the following Lagrange function:

$$L(x_1, x_2, \lambda) = x_1^2 + x_2^2 + \lambda(3x_1 - x_2 - 6) \qquad (3.27)$$

By taking the partial derivatives of (3.27) with respect to x_1, x_2, and λ and then equating to zero, respectively, we get

$$\frac{\delta L(x_1, x_2, \lambda)}{\delta x_1} = 2x_1 + 3\lambda = 0 \qquad (3.28)$$

$$\frac{\delta L(x_1, x_2, \lambda)}{\delta x_2} = 2x_2 - 1 = 0 \qquad (3.29)$$

$$\frac{\delta L(x_1, x_2, \lambda)}{\delta \lambda} = 3x_1 - x_2 - 6 = 0 \qquad (3.30)$$

Solving (3.28)–(3.30) yields

$$x_1 = \frac{13}{6}, x_2 = \frac{1}{2}, \text{ and } \lambda = \frac{-13}{9}$$

The critical point of $f(x_1, x_2)$, subject to specified condition, is $\left(\dfrac{13}{6}, \dfrac{1}{2}\right)$.

3.5.2   Linear Programming Method

This is probably the simplest and the most widely used method of optimization subject to constraints, and its history goes back to 1947 when George B. Dantzig developed the simplex method. The development of the concept of duality by J. Von Neumann in the same year helped bring about wider acceptance of the linear programming approach [7].

Past experience indicates that the application of the linear programming method has helped production and operations managers plan and make effective decisions to allocate available resources. All linear programming problems have the following properties in common [5]:

- The objective and constraint functions are expressed in terms of linear equations or inequalities.
- The problems aim to minimize or maximize some quantity (e.g., cost or profit).
- There are alternative courses of action from which to choose.
- The presence of constraints limits the persuasion of the objective.

Although in real-life environments, the objective and constraint functions could be rather complex, the simplest form of linear programming problem formulation is as follows [10]:

$$\text{Maximize (or minimize)} P = c_1 y_1 + c_2 y_2 + - - - + c_n y_n \quad (3.31)$$

Subject to

$$a_{11} y_1 + a_{12} y_2 + - - - + a_{1n} y_n (\leq,=,\geq) R_1 \quad (3.32)$$

$$a_{21} y_1 + a_{22} y_2 + - - - + a_{2n} y_n (\leq,=,\geq) R_2 \quad (3.33)$$

$$a_{m1} y_1 + a_{m2} y_2 + - - - + a_{mn} y_n (\leq,=,\geq) R_m \quad (3.34)$$

$$y_1 \geq 0 \quad (3.35)$$

$$y_2 \geq 0 \qquad\qquad (3.36)$$

$$/$$
$$/$$
$$/$$

$$y_n \geq 0 \qquad\qquad (3.37)$$

where

m	is the number of constraints.
n	is the number of variables.
P	is the profit (or cost).
y_i	is the ith variable; for $i = 1, 2, -, n$.
R_j	is the jth resource; for $j = 1, 2, -, m$.
c_i	is the ith constant; for $i = 1, 2, -, n$.
a_{ji}	is the (j, i) constant; for $j = 1, 2, -, i$ and $i = 1,2, -, n$.

Equation (3.31) is known as the objective function and (3.32)–(3.37) are known as constraints. The linear programming method is demonstrated through the following example:

Example 3.8

A company manufactures two types, (i.e., A and B) of an engineering product. The product uses three different parts, say, X, Y, and Z. In the company storage, there are 40 type-X parts, 50 type-Y parts, and 16 type-Z parts.

Each product type A uses five type-X parts and five type-Y parts. Similarly, the product type B uses four type-Z parts and four type-Y parts.

Each unit of A and B makes a profit of $30 and $15, respectively. Determine the units of A and B to be manufactured to maximize profit and compute the total maximum profit.

In this example, the total profit is expressed by

$$P = 30x_1 + 15x_2 \qquad\qquad (3.38)$$

where

P	is the total profit.
x_1	is the number of units of product type A.

x_2 is the number of units of product type B.

The available parts are used as follows:

- $5x_1 + 4x_2$ of type-Y parts;
- $5x_1$ of type-X parts;
- $4x_2$ of type-Z parts.

Using the above formulations and the given data, we write

$$\text{Maximize } P = 30x_1 + 15x_2 \tag{3.39}$$

Subject to

$$5x_1 + 4x_2 \le 50 \tag{3.40}$$

$$5x_1 \le 40 \tag{3.41}$$

$$4x_2 \le 16 \tag{3.42}$$

and

$$x_1 \ge 0 \tag{3.43}$$

$$x_2 \ge 0 \tag{3.44}$$

As this problem has only two variables, it can be solved graphically. Thus, Figure 3.3 shows plots of (3.39)–(3.44). The profit will be maximum at point D. At this point all constraints are satisfied, and it is at the extreme right with respect to profit function plots. More specifically, the profit increases to the right side.

At this point D in Figure 3.3, the values of x_1 and x_2 are 8 and 2.5, respectively. By substituting these two values into (3.39), we get

$$P = 30(8) + 15(2.5)$$
$$= \$277.5$$

In theory, it means 8 units of product type A and 2.5 units of product type B must be manufactured to maximize profit. The total maximum profit will be \$277.5.

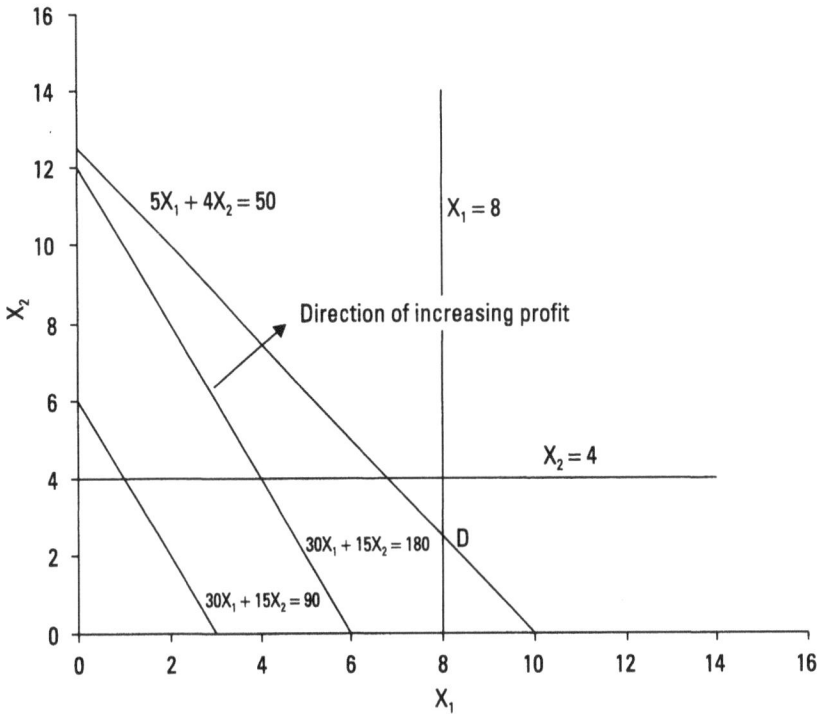

Figure 3.3 The graphical plot of Example 3.8.

3.6 Learning Curve Analysis

Past experience indicates that individuals learn by experience (i.e., get better and better at the job by carrying out the tasks more and more). This phenomenon was first reported by T. P. Wright in 1936 [11]. Nonetheless, the learning curve theory is based on assumptions such as those listed next [12]:

- The time required to complete a specified task or unit of a product or item will be less each time the task is performed.
- The unit time will reduce at a decreasing rate.
- The decrease in time will follow a certain pattern, such as negative exponential distribution shape.

The learning curves may vary from one product to another and from one organization to another. The rate of learning depends on factors such as the quality of management and the potential of the process and products [5]. Moreover, it may be said that any change in personnel, process, or product disrupts the learning curve. Consequently, there is a need for the utmost care in assuming that a learning curve is continual and permanent. Table 3.1 presents data on learning curve effects in the U.S. industrial sector [13]. Nonetheless, as per [14], an 80% learning rate is descriptive of certain operations in such areas as ship construction, electronic data processing equipment, automatic machine production, and aircraft instruments and frame assemblies.

The learning curves are found to be quite useful in a variety of applications, including strategic evaluation of company and industry performance, internal labor forecasting, establishing costs and budgets, production planning, external purchasing, and subcontracting of items [5].

The learning curve theory is based on a doubling of productivity. More specifically, when output or production doubles, the reduction in time per unit affects the learning curve rate. For example, an 80% learning rate means the second unit takes 80% of the time of the first unit, the fourth unit takes 80% of the second unit, the eighth unit takes 80% of the fourth unit, and so on. Thus, we may write [5]:

Table 3.1
Some Information on Learning Curve Effects in U.S. Industrial Sector

Number	Item/Area Description	Time Period	Cumulative Parameter	Improving Parameter	Learning Curve Slope Percentage
1	Steel making	1920–1955	Units produced (UP)	Production worker labor-hours per unit produced	79
2	Handheld calculators	1975–1978	UP	Average factory selling price	74
3	Assembly of aircrafts	1925–1957	UP	Direct labor-hours per unit	80
4	Ford Motor Company Model T production	1910–1926	UP	Price	86

$$LH_m = LH_1 m^C \qquad (3.45)$$

where

LH_m is the labor hours required to produce unit m.

LH_1 is the labor hours to produce unit one or the first unit.

C is the learning curve slope and is expressed by

$$C = \frac{\log \text{ of the learning rate}}{\log 2} \qquad (3.46)$$

The limits of the learning curve are discussed in [15].

Example 3.9

Assume that the learning rate for a certain operation is 75% and it took 90 hours to produce the first unit. Calculate the hours required to produce the fifth unit.

By substituting the given data value into (3.46), we get

$$C = \frac{\log 0.75}{\log 2}$$
$$= 0.4150$$

Using the above value and the specified data in (3.45) yields

$$LH_5 = 90(5)^{-0.4150}$$
$$= 46.15 \text{ hours}$$

It will take 46.15 hours to produce the fifth unit.

3.7 Fault Tree Analysis

This is a widely used method in industry to perform reliability analyses of engineering systems. The method was developed in the early 1960s at the Bell Laboratories to evaluate the reliability of the Minuteman Launch Control System [16]. The fault tree analysis begins by identifying an undesirable event, called the *top event,* associated with a system or problem. The events that could cause the occurrence of this top event are generated and connected by logic gates called OR, AND, and so on. A fault tree's construction proceeds by generation of fault events in a successive manner until the fault events need not be developed further.

The method is called fault tree analysis because it only considers failure or negative events, and a fault tree may simply be described as the logic structure relating the top event to the basic fault events. Figure 3.4 presents four basic symbols used in the construction of a fault tree.

The OR gate denotes that an output fault event occurs if one or more of the input fault events occur. The AND gate denotes that an output fault event occurs only if all its input fault events occur. The circle represents a basic fault event (e.g., the failure of an elementary component). The rectangle denotes an event that occurs due to combination of fault events through the input of a logic gate such as OR and AND. A comprehensive list of fault tree symbols is given in [16]. The probability of occurrence of the OR gate output fault event is

$$P_{OR} = 1 - \prod_{i=1}^{n}(1 - P_i) \qquad (3.47)$$

where

P_{OR} is the occurrence probability of the OR gate output fault event.

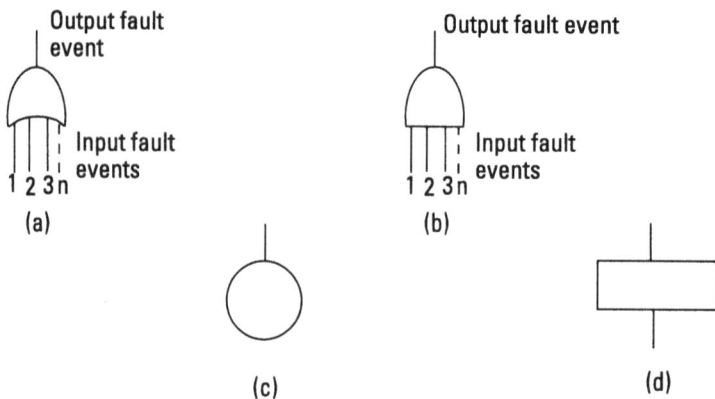

Figure 3.4 Basic fault tree symbols: (a) OR gate, (b) AND gate, (c) basic fault event, and (d) resultant event.

n is the number of gate input fault events.

P_i is the occurrence probability of the input fault event i, for $i = 1, 2, -, n$.

Similarly, the probability of occurrence of the AND gate output fault event is

$$P_{AND} = \prod_{i=1}^{n}(1 - P_i) \qquad (3.48)$$

The fault tree analysis concept can also be applied in making various types of management-related decisions. The following example demonstrates the application of the fault tree analysis method to a management problem.

Example 3.10

Assume that an engineering company management wishes to determine the probability of failure of completing a product-development project. The project completion can be unsuccessful due to any of the three basic factors: A (unavailability of trained manpower), B (unavailability of the amount of money required), or C (failure to overcome technical problems). The required trained manpower could be unavailable due to either factor: a_1 (stiff competition), or a_2 (poor company image). Two important factors for the unavailability of the amount of money required are either b_1 (poor initial estimate), or b_2 (unexpected rise in the cost of services and resources). There are two possible reasons for poor company image: a_{21} (poor labor relations), or a_{22} (poor salaries and benefits). Develop a fault tree for the top event, X, that the product-development project will not be completed successfully. Also, calculate the probability of occurrence of event X if the probabilities of occurrence of events $C = 0.1$, $b_1 = 0.05$, $b_2 = 0.08$, $a_1 = 0.04$, $a_{21} = 0.01$, and $a_{22} = 0.07$, and all events occur independently.

Using the symbols in Figure 3.4, the fault tree of Figure 3.5 for Example 3.10 was developed.

Using the specified data in (3.47), the probability of the occurrence of event a_2 is

$$P_{a_2} = 1 - (1 - 0.01)(1 - 0.07)$$
$$= 0.0793$$

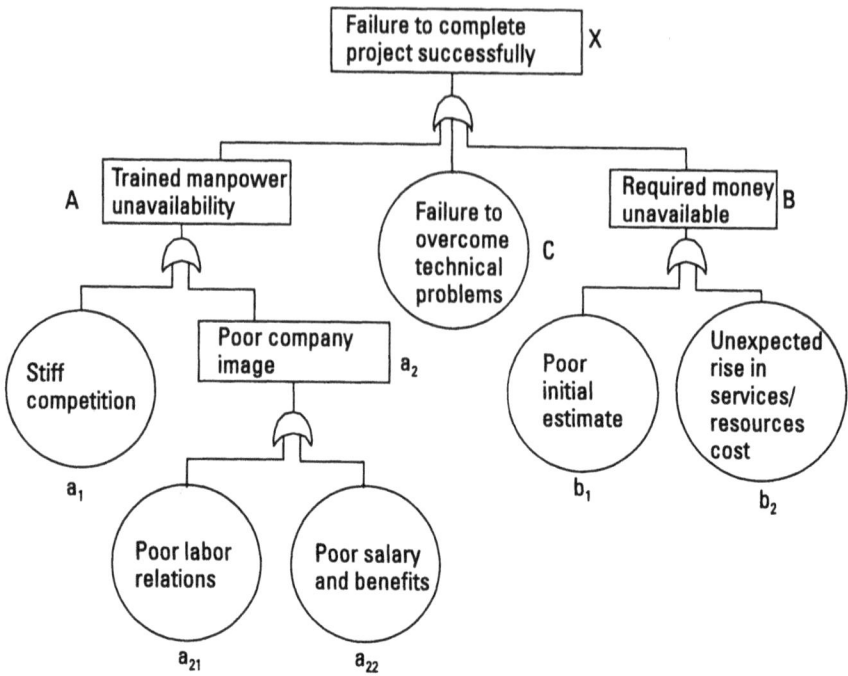

Figure 3.5 A fault tree for Example 3.10.

Using the above result and the given data in (3.47), the probability of the occurrence of event A is given by

$$P_A = 1 - (1 - 0.04)(1 - 0.0793)$$
$$= 0.1161$$

By inserting the given data values into (3.47), the occurrence probability of event B is

$$P_B = 1 - (1 - 0.05)(1 - 0.08)$$
$$= 0.1260$$

Using the calculated values and the given data in (3.47), the probability of occurrence of event X is given by

$$P_X = 1-(1-0.1161)(1-0.1)(1-0.1260)$$
$$= 0.3047$$

The fault tree in Figure 3.5 is redrawn in Figure 3.6, and it shows the given and the earlier calculated probability values. Thus, the probability of occurrence of the top event X (i.e., the product-development project will not be successfully completed) is 0.3047.

3.8 Forecasting

Each day engineering managers make various decisions that are concerned with some point in the future. An example is the decision to produce a certain number of units of a product. One important input in making such a decision is the demand for the product. Forecasting is a useful tool to estimate demand for a certain product, and it may simply be described as the art and science of predicting future events [5, 17].

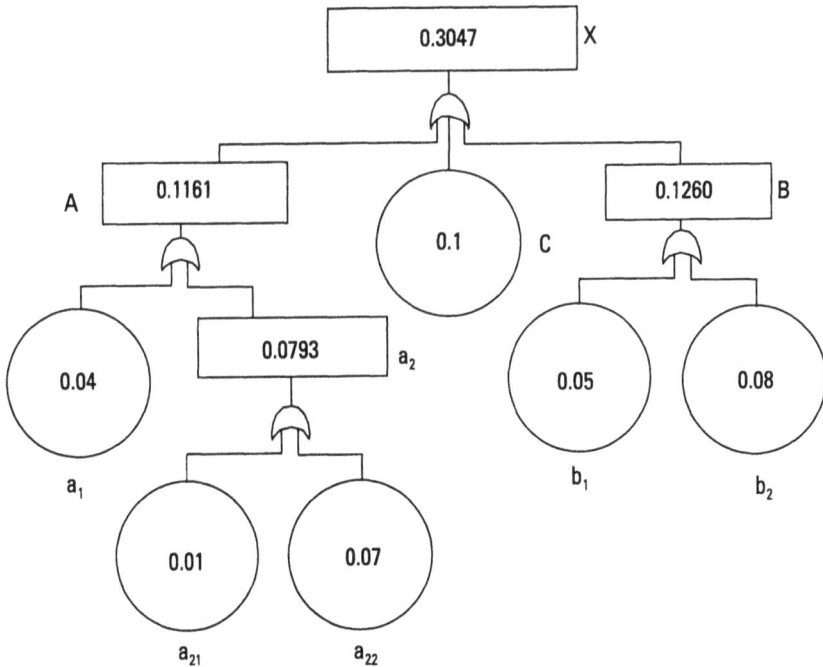

Figure 3.6 A probability fault tree for Example 3.10.

3.8.1   Forecasting Types and Time Horizons

There are essentially three major types of forecasts used by companies in planning the future of their operations, as shown in Figure 3.7 [5]. The technological forecasts are concerned with predicting the rates of technological progress that may lead to the development of new products requiring new plants and equipment.

The demand forecasts are concerned with projecting demand for an organization's products or services that in turn drive company's production, capacity, and scheduling systems, and serve as inputs to areas such as marketing, finance, and personnel planning. Economic forecasts are concerned with predicting factors such as inflation rates, money supplies, housing starts, and other planning indicators.

Usually, forecasts are categorized by the future time horizon they cover, and their three classifications are shown in Figure 3.8 [18]. The long-range forecast usually covers a period of 3 or more years and is used in areas such as planning for new products, facility location or expansion, research and development, and capital expenditures. The short-range forecast covers a time span of up to 1 year, but often it is less than 3 months. It is used in areas such as purchasing, planning, job assignments and scheduling, and levels of work force. The medium-range forecast usually covers a time period from 3

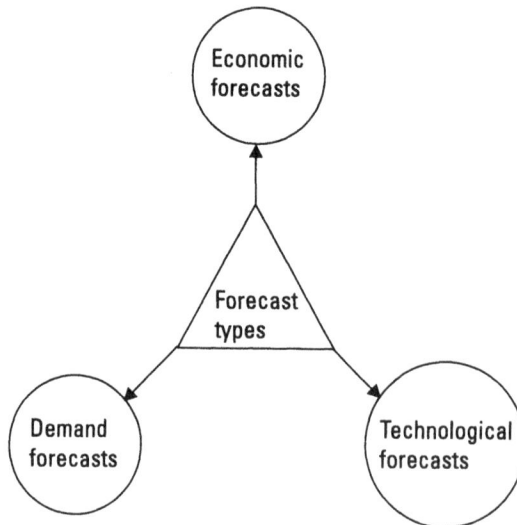

Figure 3.7 Types of forecasts.

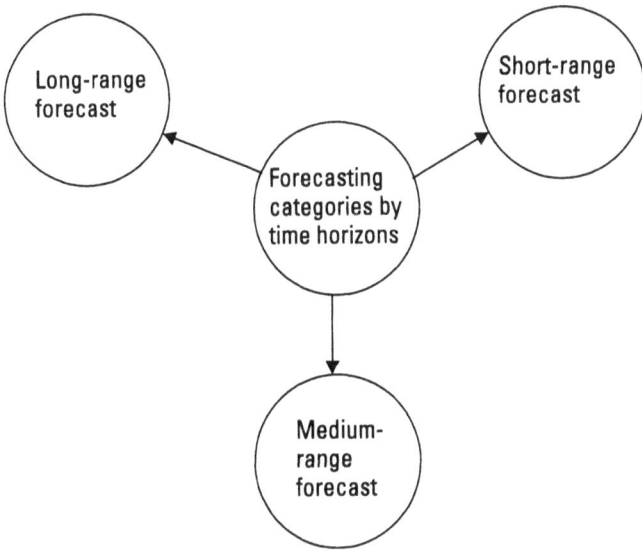

Figure 3.8 Forecasting categories by time horizons.

months to 3 years and finds applications in many areas including sales planning, budgeting, and production planning.

3.8.2 Forecasting Steps, Methods, and Technique Selection Factors

The success of any forecast depends on the effectiveness of the approach followed in forecasting. A useful forecasting approach is composed of the following eight steps [5]:

- Determine the forecast application and objective.
- Choose with care the items to be forecasted.
- Determine forecast time horizon (i.e., long, short, or medium).
- Choose appropriate forecasting model(s).
- Collect the appropriate data required to make the forecast under consideration.
- Validate the forecasting model with care.
- Make all relevant forecasts.
- Implement the appropriate results.

The forecasting methods may be divided into two broad categories: qualitative and quantitative. The qualitative methods provide forecasts that incorporate factors such as the decision maker's emotions, personal experiences, and intuition. Some examples of the qualitative methods are jury of executive opinion, Delphi method, consumer market survey, and sales force opinion composite [5]. The quantitative methods provide forecasts that were obtained by employing various mathematical models that use past data or causal variables to forecast demand. Examples of the quantitative methods include exponential smoothing, moving averages, linear-regression causal model, and trend projection.

The forecasting technique selection could be subject to one or more factors, such as the forecast development cost, the availability of historical data, the length of prediction interval, data accuracy, the time for analysis, the expected accuracy from the forecasted result, and the complexity of factors affecting operations in time to come [14]. Two widely used quantitative methods are presented next.

3.8.3   Simple Average

This is a simple and straightforward method in which the demands of all previous periods are given equal weight. Thus, the simple average of the past is expressed by [7]

$$SA = \sum_{i=1}^{m} D_i / m \qquad (3.49)$$

where

SA is the simple average.

m is the number of all the past demand periods.

D_i is the demand for the ith previous past period.

The value of SA will be the forecasted value of demand for future periods. The main advantage of this forecasting approach is that the effects of randomness are minimized because it uses all of the data of past periods in the calculation.

Example 3.11

Assume that in the past four 5-month periods, the demand for a certain company product was 200, 150, 100, and 250 units, respectively. Determine the simple average of demands.

By substituting the specified data into (3.49), we get

$$SA = \frac{200 + 150 + 100 + 250}{4}$$

$$= 175 \text{ units}$$

It means a forecast for demand for a future 5-month period is 175 units.

3.8.4 Exponential Smoothing

This is a frequently used and sophisticated weighted moving-average forecasting method. The method is fairly easy to use and requires very little record keeping of past data.

The exponential smoothing approach forecasts for one period ahead by weighing the most recent past period data or demand most heavily. In a continuing manner, the weights assigned to successively past period data or demands decrease exponentially. Thus, the method is called exponential smoothing.

The basic exponential smoothing formula is as follows:

$$F_a = \alpha D_{a-1} + (1 - \alpha) F_{a-1}, \text{ for } 0 \le \alpha \le 1 \qquad (3.50)$$

where

a	is the time period.
F_a	is the forecast for demand one period ahead.
α	is the smoothing constant or the weighting factor.
F_{a-1}	is the demand forecast for the most recent past period.
D_{a-1}	is the actual demand for the most recent past period.

Using (3.50) we write down the following expression for the forecast for the period just ending:

$$F_{a-1} = \alpha D_{a-2} + (1 - \alpha) F_{a-2} \qquad (3.51)$$

where

D_{a-2}	is the actual demand for the period $(a - 2)$.

F_{a-2} is the forecast for the period $(a-2)$.

In a similar manner, we write down the following equation for the period $(a-2)$ forecast:

$$F_{a-2} = \alpha D_{a-3} + (1-\alpha)F_{a-3} \qquad (3.52)$$

where

D_{a-3} is the actual demand for the period $(a-3)$.
F_{a-3} is the forecast for the period $(a-3)$.

By inserting (3.52) into (3.51), we get

$$F_{a-1} = \alpha D_{a-2} + \alpha(1-\alpha)D_{a-3} + (1-\alpha)^2 F_{a-3} \qquad (3.53)$$

Substituting (3.53) into (3.50) yields

$$F_{a} = \alpha D_{a-1} + \alpha(1-\alpha)D_{a-2} + \alpha(1-\alpha)^2 D_{a-3} + \alpha(1-\alpha)^3 F_{a-3} \qquad (3.54)$$

The generalized version of (3.54) is as follows:

$$F_{a} = \alpha D_{a-1} + \alpha(1-\alpha)D_{a-2} + \alpha(1-\alpha)^2 D_{a-3}$$
$$+ - - - + \alpha(1-\alpha)^{n-1} F_{a-n} + (1-x)^n F_{a-n} \qquad (3.55)$$

where n is the total number of past periods.

The weights in the right hand side of (3.55) decrease by factor $(1-\alpha)$ or more simply exponentially. This means that the more recent data are given greater weight than the remote data.

For a very low value of α in (3.55) all past data or observations receive almost equal weights. In contrast for $\alpha = 1$, the demand or data of the last period is the forecast for the period ahead. More specifically, 100% weight is given to the demand or data of the last period. The past experience indicates that the frequent used values for α are between 0.01 and 0.3 [7].

Example 3.12

Assume that a computer manufacturer sold 500, 400, 300, 500, 350, and 400 personal computers for the months of January, February, March, April,

May, and June, respectively. The forecast for January was 600 personal computers to be sold. Using (3.55), forecast the number of personal computers to be sold by the manufacturer for the month of July, if $\alpha = 0.3$.

Substituting the given data values into (3.55) yields

$$F_a = (0.3)\,400 + (0.3)(1-0.3)\,350 + (0.3)(1-0.3)^2\,500$$
$$+ (0.3)(1-0.3)^3\,300 + (0.3)(1-0.3)^4\,400 + (0.3)(1-0.3)^5\,500$$
$$+ (1-0.3)^6\,600 \approx 422 \text{ personal computers}$$

It means that the manufacturer should expect to sell 422 personal computers for the month of July.

3.9 Problems

1. An engineering manufacturer sold 80, 90, 70, 100, and 110 electric motors for the months of July, August, September, October, and November, respectively. The forecast for July was 95 electric motors to be sold. Using the exponential smoothing method, forecast the number of electric motors to be sold by the manufacturer for the month of December, if the value of the smoothing constant is 0.2.

2. What are the types of forecasts? Discuss each in detail.

3. Discuss the following terms:

   - Compound interest;
   - Present value;
   - Learning curve;

4. Prove that the present value of a single payment is given by

$$PV = A_m (1+i)^{-m} \qquad\qquad (3.56)$$

where

PV	is the single payment present value.
m	is the number of interest periods or years.
i	is the compound interest rate per period.
A_m	is the compound amount at the end of the mth period or year.

5. Describe the following two methods of depreciation:

- The sum-of-the-years-digits method;
- The declining-balance method.

6. What are the steps involved in analyzing problems with decision trees?

7. Describe the following:

- A decision tree;
- A fault tree;
- Linear programming.

8. Assume that an engineering company manufactures type X and type Y of a product. Both types use three different materials, say, A, B, and C. The company has 30, 40, and 20 tons of materials A, B, and C, respectively. Each unit of type X uses 2 tons of A material and 3 tons of B material. Similarly, the type Y unit uses 4 tons of C material and 3 tons of B material. Each unit of X and Y makes a profit of $20,000 and $30,000, respectively. Determine the units of X and Y to be manufactured to maximize profit and calculate the total maximum profit.

9. Assume that the learning rate for a specific operation is 80% and it took 150 hours to produce the first unit. Calculate the hours needed to produce the fourth unit by using (3.45).

10. Discuss two broad categories of forecasting methods.

References

[1] Howell, J. E., and D. Teichroew, *Mathematical Analysis for Business Decisions*, Chicago, IL: Irwin, 1971.

[2] Dhillon, B. S., *Life Cycle Costing: Techniques, Models, and Applications*, New York: Gordon and Breach Science Publishers, 1989.

[3] Riggs, J. L., *Economic Decision Models for Engineers and Managers*, New York: McGraw-Hill, 1968.

[4] Lee, S. M., L. J. Moore, and B. W. Taylor, *Management Science*, Dubuque, IA: Wm. C. Brown Company Publisher, 1981.

[5] Heizer, J., and B. Render, *Production and Operations Management*, Upper Saddle River, NJ: Prentice Hall, 1996.

[6] Tellier, R. D., *Operations Management*, New York: Harper & Row, 1978.

[7] Dhillon, B. S., *Engineering Management*, Lancaster, PA: Technomic Publishing Company, 1987.

[8] Magee, J. F., "Decision Trees for Decision Making," *Harvard Business Review*, July/August 1964, pp. 126–138.

[9] Gue, R. L., and M. E. Thomas, *Mathematical Methods in Operations Research*, London: Macmillan, 1968.

[10] Dantzig, G. B., *Linear Programming and Extensions*, Princeton, NJ: Princeton University Press, 1963.

[11] Wright, T. P., "Factors Affecting the Cost of Airplanes," *Journal of the Aeronautical Sciences*, February 1936, pp. 28–34.

[12] Chase, R. B., and N. J. Aquilano, *Production and Operations Management*, Chicago, IL: Irwin, 1981.

[13] Cunningham, J. A., "Using the Learning Curve as a Management Tool," *IEEE Spectrum*, June 1980, pp. 45–46.

[14] Riggs, J. L., *Production Systems: Planning, Analysis, and Control*, New York: John Wiley and Sons, 1981.

[15] Abmathy, W. J., and K. Wayne, "Limits of the Learning Curve," *Harvard Business Review*, Vol. 52, September/October 1974, pp. 109–119.

[16] Dhillon, B. S., and C. Singh, *Engineering Reliability: New Techniques and Applications*, New York: John Wiley and Sons, 1981.

[17] Georgoff, D. M., and R. G. Murdick, "Managers Guide to Forecasting," *Harvard Business Review*, Vol. 64, January/February 1986, pp. 110–120.

[18] Stonebraker, P. W., and G. K. Leong, *Operations Strategy*, Boston, MA: Allyn and Bacon, 1994.

4

Project Selection and Management

4.1 Introduction

Project management has emerged because characteristics of our late twentieth-century society demanded the development of new methods and techniques of management, but its roots may be traced back to ancient times. For example, project management was used to a certain degree in constructing historical architectural wonders such as Roman buildings and roads [1].

The origin of the modern project management may be traced back to the late 1950s when Paul O. Gaddis published an article entitled "The Project Manager" in the *Harvard Business Review* [2]. This paper discussed the role of the project manager in an advanced-technology industrial sector, the prerequisites for carrying out the project-management task, and the type of training necessary to develop an individual to manage projects. In 1969, the Project Management Institute (PMI) was established, and it has played an instrumental role in the development of the project management field [3]. Today, PMI has around 10,000 members. Since the late 1950s, more than 150 books and hundreds of articles have appeared on various aspects of project management [1].

Project selection may simply be described as the process of evaluating projects, and then choosing to implement some of them to achieve the goals of the parent organization. Thus, project selection is very important to engineering companies, because through the evaluation process, intelligent decisions concerning investment in projects can be made. More specifically, these decisions are crucial to the success or failure of the companies.

There are four objectives of project selection: choose only the viable projects, reject the undesirable projects, ensure that there is no damaging effect on flow of project proposals, and ensure that there is no rejection of desirable projects [4].

This chapter presents various different aspects of project selection and management.

4.2   Terms and Definitions

This section presents terms and definitions directly or indirectly related to project selection and management [1, 5, 6].

- *Project.* This is a plan of work job assignment, or task. (It is also referred to as a job or task.)
- *Project management.* This is the art of directing and coordinating material and human resources throughout the project life span by utilizing various management methods and techniques to achieve effectively predetermined goals of scope, quality, time, cost, and participant satisfaction [7].
- *Project manager.* This is a person with an assigned direct responsibility for project execution.
- *Project organization.* This is an establishment of people, procedures, responsibilities, and authority.
- *Project selection.* This is the process of evaluating projects and then choosing to implement some of them to achieve the goals of the parent organization.
- *Project objectives.* These are the goals, as measured by time parameters, constraints or targets, results, milestones, and control considerations, and usual project sequencing.
- *Project team.* This consists of all participants involved in the project (i.e., full or part time).
- *Project parameters.* These are the factors, specifications, or characteristics of the complete project or of its parts.
- *Project proposal.* This is a summary of a proposed project concerning its managerial, schedule, and technical contents.
- *Project control report.* This is a management report and an operations tool utilized to report manpower- or financial-related information in a summarized form both by program and organization.

4.3 Types of Information Required for Evaluating a Project

Usually, various types of information are needed to evaluate a project. Figure 4.1 presents five main categories of such information [3].

Tables 4.1–4.5 present important subelements of required marketing-related, production-related, manpower-related, finance-related, and administrative-related and miscellaneous information, respectively. It is to be noted that these subelements may vary from one project to another.

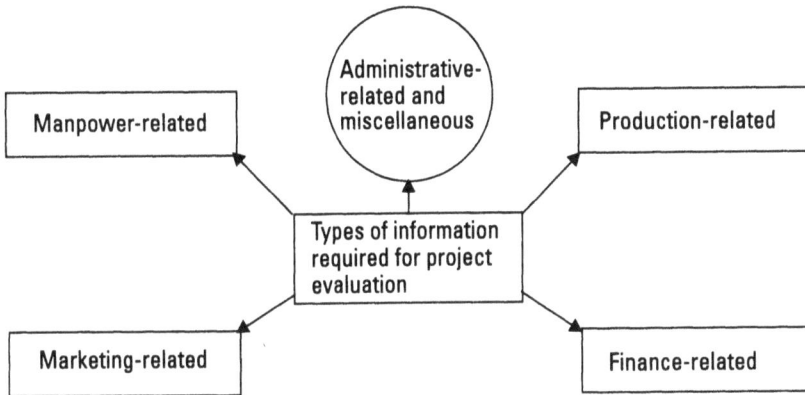

Figure 4.1 Main categories of information required for project evaluation.

Table 4.1
Important Subelements of Required Marketing-Related Information

Number	Subelement
1	Acceptance by consumer
2	Potential market and its size
3	Impact on consumer safety
4	Probable market share
5	Possibility for spin-off
6	Impact on existing product line
7	Ability to control quality
8	Time until acquiring market share
9	Output's estimated life

Table 4.2
Important Subelements of Required Production-Related Information

Number	Subelement
1	Availability of raw materials
2	Energy needs
3	Equipment and facility requirements
4	Process safety
5	Degree and length of disruption during installation
6	Effect on waste
7	Compatibility with current technological know-how
8	Change in quality of output and quality-control procedures
9	Required development time
10	Time until ready to install
11	Impact on suppliers
12	Other areas of applications of technology
13	Change in raw material usage, time to produce a unit, and cost to produce a unit

Table 4.3
Important Subelements of Required Manpower-Related Information

Number	Subelement
1	Availability of required skilled labor
2	Skill and training requirements
3	Impact on current work conditions
4	Change in work force size
5	Degree of resistance from existing labor force
6	Need for support labor
7	Needs for intergroup communication

Table 4.4
Important Subelements of Required Finance-Related Information

Number	Subelement
1	Investment required
2	Profit margin
3	Payout period
4	Degree of financial risk
5	Impact on cash flows
6	Break-even time
7	Impact on cyclical and seasonal fluctuations

Table 4.5
Important Subelements of Required Administrative-Related and Miscellaneous Information

Number	Subelement
1	Impact on information system and computer usage
2	Patent or trade secret protection
3	Degree of impact on image with suppliers, competitors, and customers
4	Meeting of government safety and environmental standards
5	Degree of consulting help required from inside and outside the organization
6	Usefulness of new process
7	Vulnerability to single consumer or supplier
8	Cost to maintain skill in new technology
9	Capacity of management to direct and control new process
10	Degree of understanding of new technology
11	Difference between the new and current process
12	Stockholders' reactions

4.4 Project Selection Models

Over the years, a large number of mathematical models for selecting projects have been developed. Some of these models are actually being used to select projects in the industrial sector. For example, a survey of 40 high-level staff members from 29 Fortune 500 companies revealed that 80% of the respondents reported using one or more financial models in research and development–related project decision making [3, 8]. Nonetheless, often reasons such as those listed next are cited for rarely using the mathematical models in project selection in industry or government [9–11]:

- Inadequate reliable estimates for cost, performance, utility, and time;
- Inadequate recognition of the interrelationships between projects;
- Dynamic nature of project selection and review;
- Poor handling of uncertainty and risk;
- Poor consideration given to the relationship between the effort applied to a project and the chance for success;
- Necessity for multiple criteria for evaluation.

Past experience indicates that in choosing a mathematical model for use in project selection, considerations must be given to factors such as realism, ease of use, flexibility, capability, cost, and easy computerization [3, 12]. This section presents a number of mathematical models that can be useful in making project selection decisions [3, 11, 13–19].

4.4.1 Benefit-Cost Ratio Model

This simple model is concerned with performing analysis to determine if benefits from a given project outweigh its cost. More specifically, it is only worthwhile to consider a project for development if its benefits are far greater than the investments. Nonetheless, the benefit/cost ratio is defined by [13]

$$BCR = \frac{UB}{TIC} \tag{4.1}$$

where

BCR is the benefit/cost ratio.

TIC is the total investment cost including the operating cost.

UB is the user benefits.

For a project to be beneficial, the value of BCR must be greater than unity.

Example 4.1

Assume that a machine's acquisition cost is $100,000 and its expected useful life is 15 years. The annual maintenance plus operation cost of the machine is estimated to be $4,000. The lifetime benefits of the machine are estimated to be $200,000. Calculate the value of the benefit/cost ratio.

The total investment cost of the machine is:

$$TIC = 100,000 + (4,000)(15) = \$160,000$$

Substituting the above and specified values into (4.1) yields

$$BCR = \frac{200,000}{160,000}$$
$$= 1.25$$

The value of the benefit/cost ratio is 1.25.

4.4.2 Disman Model

This is a useful model to determine the maximum amount of money justifiable for a project. This model was developed by S. Disman. He reasoned that the income over a given period from an investment is [20]

$$I = INV + i.INV \tag{4.2}$$

where

I is the income from an investment over a specified period.

INV is the amount of investment.

i is the rate of return on the investment.

For a 1-year period, from (4.2), we get

$$INV = \frac{I}{(1+i)}$$

The above equation is used to estimate the maximum investment justifiable for a project when the rate of return and the annual net income are known. Thus, for 1 year, we write

$$INV_{m1} = \frac{NI}{(1+i)} \tag{4.3}$$

where

    INV_{m1} is the maximum investment justifiable for 1 year.
    NI is the annual net income.

For n years, using (4.3) we write

$$INV_{mn} = \sum_{j=1}^{n} \frac{NI_j}{(1+i)^j} \tag{4.4}$$

where

    INV_{mn} is the maximum investment justifiable for n years.
    NI_j is the net income from the project for year j.

Equation (4.4) is modified to the following form to take into consideration the risks associated with the technical and commercial successes of the project or product:

$$INV_{max} = P_t P_c \sum_{j=1}^{n} \frac{NI_j}{(1+i)^j} \tag{4.5}$$

where

    P_c is the risk of project or product's commercial success.
    P_t is the risk of project or product's technical success.
    INV_{max} is the maximum investment justifiable for n years.

Example 4.2

A company anticipates developing a new machine with an expected life of only 3 years. The expected annual net income from the machine for 3 years is $80,000. The estimated risks of machine's technical success and commercial success are 0.8 and 0.7, respectively. Calculate the maximum investment justifiable for the new machine, if the rate of return on the investment is estimated to be 40%.

Substituting the specified data into (4.5) yields

$$INV_{max} = (0.8)(0.7)\left[\frac{80,000}{(1+0.4)} + \frac{80,000}{(1+0.4)^2} + \frac{80,000}{(1+0.4)^3}\right]$$

$$= \$71,183.67$$

The maximum investment justifiable for the new machine is $71,183.67.

4.4.3 Pacifico Model

This is the profitability index of acceptability, or more simply the benefit/cost ratio. A project can be accepted, if the value of this ratio is greater than unity. The equation for the model is expressed by [3]

$$PIA = \left[\left(\prod_{i=1}^{4} P_i\right)(AS)(AP)\sqrt{L}\right]/IC \qquad (4.6)$$

where

PIA	is the profitability index of acceptability.
P_i	is the probability of i; $i = 1$ (research success), $i = 2$ (commercial success, given process success), $i = 3$ (development success, given research success), and $i = 4$ (process success, given development success).
AP	is the average annual profit per unit.
AS	is the average annual sales volume in units of products.
L	is the expected life of the product extension expressed in years.
IC	is the investment or the total cost of the research and development effort for the project under consideration.

4.4.4   Mottley and Newton Model

This model computes an overall score for each project and then selects the project with the highest score for development. The model is based on five basic factors: project cost, product or project completion time, strategic need, promise of success, and market gain [21]. All of these factors are assumed to contribute to the potential profits of the company.

The equation for the model is expressed by

$$OS = \prod_{i=j}^{5} RS_i \tag{4.7}$$

where

OS is the overall score of a project.

R_{si} is the rated score of the ith factor; i = 1 (the strategic need), i = 2 (product or project completion time), i = 3 (project cost), i = 4 (promise of success), i = 5 (market gain).

Each of the above five factors is assigned a score from 1 to 3 and reasons for assigning scores are tabulated in [21, 22].

Example 4.3

An engineering company is proposed with four product development projects: A, B, C, and D, and it would like to develop only one of these projects. Table 4.6 presents scores for five factors associated with each of these projects. Determine the most promising project for development.

Table 4.6
Scores for Proposed Product Development Projects

Number	Factor Description	Project A	Project B	Project C	Project D
1	Strategic need	2	1	1	1
2	Product or project completion time	3	3	3	3
3	Project cost	2	3	2	3
4	Promise of success	2	2	1	1
5	Market gain	1	2	3	1

Using the given data in Table 4.6 for Project A in (4.7), we get

$$OS = (2)(3)(2)(2)(1)$$
$$= 24$$

Substituting the given data in Table 4.6 for Project B into (4.7) yields

$$OS = (1)(3)(3)(2)(2)$$
$$= 36$$

Using the given data in Table 4.6 for Project C in (4.7), we get

$$OS = (1)(3)(2)(1)(3)$$
$$= 18$$

Inserting the given data in Table 4.6 for Project D into (4.7) yields

$$OS = (1)(3)(3)(1)(1)$$
$$= 9$$

By examining the above four results, Project B has the highest score. Thus, it qualifies as the most promising project for development.

4.4.5 Calculated-Risk Model

This model is the simplified version of return on investment approach and, for screening purposes, it is extremely useful for comparing with a predetermined minimum rate. The equation for the model is expressed by [23]

$$I = \frac{(TNAI)RF}{TE} \tag{4.8}$$

where

    I is the index.

    $TNAI$ is the total net annual incomes from the project.

    RF is the risk factor that represents all risks associated with incomes.

TE   is the total expenditure that includes capital assets, product, and market development.

4.4.6 Sobelman Model

This model takes into consideration the time value of the money by determining the present value of average profits and costs. The project is considered for development only when its index value is positive. The equation for the model is defined by [24]

$$VP = AANP \sum_{j=1}^{N} \frac{1}{(1+i)^j} - AACPD \sum_{j=1}^{M} \frac{1}{(1+i)^j} \qquad (4.9)$$

$$N = N_1 + N_2 \left(1 - \frac{M_1}{M_2}\right) \qquad (4.10)$$

$$M = M_1 + M_2 \left(1 - \frac{N_1}{N_2}\right) \qquad (4.11)$$

where

VP   is the value of project.

AACPD   is the average annual cost of the product development.

AANP   is the average annual net profit.

i is the interest rate.

M_1 is the product completion time expressed in years.

M_2 is the group of those products' average development time (expressed in years) of which the project under consideration is a member.

N_1 is the product useful life expressed in years.

N_2 is the group of those products' average useful life (expressed in years) of which the project under consideration is a member.

Example 4.4

Assume that we have the following data: N_1 = 8 years, M_1 = 3 years, i = 0.05, N_2 = 4 years, M_2 = 2 years, AANP = $30,000, and AACPD = $5,000. Using the above data in (4.9)–(4.11), calculate the value of VP.

By substituting the given data into (4.9)–(4.11), we get

$$M = 3 + 2\left(1 - \frac{8}{4}\right) = 1$$

$$N = 8 + 4\left(1 - \frac{3}{2}\right) = 6$$

$$VP = (30,000)\sum_{j=1}^{6}\frac{1}{(1+0.05)^{j}} - (5,000)\sum_{j=1}^{1}\frac{1}{(1+0.05)^{j}}$$

$$= (30,000)(5.0757) - (5,000)(0.9524)$$

$$= \$147,508.76$$

The value of the project is $147,508.76.

4.4.7 Manley Model

This model can also be used to evaluate new projects, and its basis is the net profit/sales ratio. The model ratio is defined by [22]

$$MR = \frac{0.5\,RC + WC + E}{(ATS)\theta} \tag{4.12}$$

where

θ	is the time to recover total investment expressed in years.
ATS	is the annual total sales of the product.
RC	is the research cost prior to taxes.
WC	is the working capital.
E	is the expenditure in plant.

4.4.8 Profitability Index Model

This index is based on constant demand and is expressed by [22]

$$I = \frac{P_{ts}P_{cs}(USP)(SML)(AASV)}{TE} \tag{4.13}$$

where

I	is the index.
TE	is the total expenditure on the project.
$AASV$	is the annual average sales volume.
SML	is the static market life.
USP	is the unit selling price.
P_{cs}	is the product's commercial success probability.
P_{ts}	is the product's technical success probability.

4.4.9 Relative Worth Index Model

This is another model used to select projects and is known as the index of relative worth. The index is defined by [22]

$$RWI = [P_{ts}P_{csd} + P_{se}A] / I_{rd} \qquad (4.14)$$

$$A = \sum_{k=0}^{n} \frac{NI_k}{(1+i)^k} - \sum_{k=0}^{n} \frac{I_k}{(1+i)^k} \qquad (4.15)$$

where

i	is the interest rate.
I_{rd}	is the investment for research and development of product.
P_{se}	is the success probability of the product because of economic conditions.
P_{csd}	is the commercial success probability of the product because of new design.
P_{ts}	is the technical success probability of the product.
n	is the market life of the product expressed in years.
I_k	is the investment in year k.
NI_k	is the net income in year k.
RWI	is the index of relative worth.

4.4.10 Dean-Sengupta Model

This is another model used in selecting projects, and it becomes possible to obtain information on the relative performance of two projects by comparing their respective indexes. The model index is expressed by [25]

$$I_{ds} = P_{ts} P_m R \, X/Y \, . \, AIC \quad\quad (4.16)$$

$$X = \sum_{j=1}^{M} \frac{1}{(1+i)^j} \quad\quad (4.17)$$

$$Y = \sum_{j=0}^{M} \frac{1}{(1+i)^j} \quad\quad (4.18)$$

where

I_{ds} is the Dean-Sengupta index.

P_{ts} is the project's technical success probability.

P_m is the probability that the product will be marketed given it is successful technically.

R is the annual return from a project, if it is a technical success and marketed.

AIC is the annual investment cost.

i is the annual interest rate.

M is the project's useful life expressed in years.

4.4.11 Net Income Present Value Model

This is a useful model to determine the present value of net income from a project. The equation for the model is expressed by [11]

$$PW = i + \frac{NI}{AVNRI} \frac{1}{(ANVRI + 1)^{n-1}} \quad\quad (4.19)$$

where

PW is the net income present value.

$AVNRI$ is the average value of net return on investment.

NI is the net income.

i is the interest rate.

n is the time to recover capital expressed in years.

4.5 Need for Project Management, Project Organization Life Cycle Phases, and Project Management Functions and Procedure Characteristics

Although many factors require the practice of project management, some of the frequently cited factors are the meeting of a project's tight time schedules, the close controlling of a project due to penalty contracts, the disruption of an organization's ongoing activities with the integration of project activities, and requirements for major innovations [22].

A project organization may be divided into the following phases [13, 22]:

- *Investigation phase.* This is the first phase of project organization and tasks such as developing specifications, identifying principal problems, and grouping principal tasks are performed during this phase.

- *Start-up phase.* This phase begins after the management commits to a project and appoints a project manager.

- *Growth phase.* This phase is concerned with building the project organization.

- *Maturity phase.* This is the fourth phase of project organization and, during the phase, the project organization's strength peaks.

- *Decline phase.* This phase begins soon after solving the principal design problems.

- *Phase-out phase.* Usually, the project is phased out soon after the buyer officially accepts the delivery of the product.

Project management performs many functions, and they may be categorized into seven major areas as shown in Figure 4.2 [25].

The characteristics of a project management procedure include limited life project, dynamic changes in responsibilities and organization, project-oriented organization, frequent progress evaluation, poor project management control over participants from other firms, and clearly defined goal, task, and subtask responsibilities [13, 22].

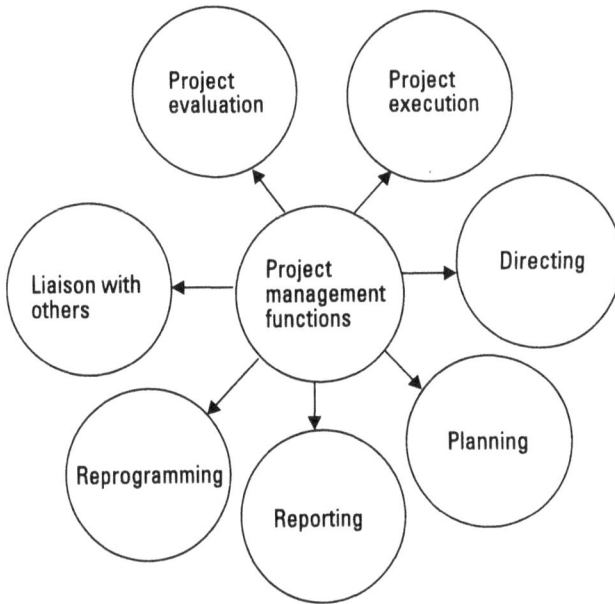

Figure 4.2 Project management functions.

4.6 Project Manager Responsibilities, Qualifications, Selection, and Reporting

Although a project manager performs various types of functions, his or her three basic responsibilities are as follows [13]:

- To ensure that the product under consideration is produced within the allocated budget;
- To ensure that the product under consideration is delivered to the customer within the limits of a defined time schedule;
- To ensure that the product under consideration satisfies fully all of the specified performance requirements.

A successful project manager possesses qualifications such as [2, 13]:

- Appropriate previous experience;
- Thorough understanding of profitability concepts;

- Effective knowledge of management problems (e.g., marketing, contracts, personnel administration, control and purchasing, and law);

- A burning desire to train and develop supervisory personnel;

- Sufficient working knowledge in various areas of science.

In selecting a project manager, factors such as strong technical background and experience, the ability to get along well with senior management, previous experience working in various departments, being a good team player, being currently available, being a mature and a hard-nosed individual, and walking on water must be considered seriously [3].

The reporting of the project manager within the company organization must be carefully considered. Some important reasons for having a project manager report directly to a high level in the organizational setup are as follows [25, 26]:

- To provide the project manager sufficient status to perform his or her job effectively;

- To impress customers, particularly in a competitive environment;

- To obtain adequate and timely assistance in solving problems;

- To report to the individual who directs functions from which the project manager is expected to get results through coordination.

By contrast, some of the reasons for having the project manager reporting to a lower level of management are as follows [25, 26]:

- Too many projects, in particular, the small ones, reporting to senior management leads to organizational and operational inefficiencies.

- Placing a junior project manager too high up on the management ladder may alienate senior functional executives on whom the project manager relies for support.

- Putting a small project high up on the management ladder may generate an illusion of executive attention, but in reality it fosters executive neglect of the project.

4.7 Project Management Methods

Two widely used methods for planning and controlling projects are known as the *critical path method* (CPM) and *program evaluation and review technique* (PERT). The history of the CPM may be traced back to 1956, when E. I. du Pont de Nemours and Company used a network model for scheduling design- and construction-related activities [27]. The PERT was developed about same time as CPM by a team formed by the U.S. Navy's Special Project Office in 1958. The team included members from the Lockheed Missile System Division and from Booz, Allen, and Hamilton, a consulting company [27].

Usually, the three important factors of concern in a given project are time, cost, and availability of resources, and both (CPM and PERT) can easily handle such factors individually and in combination. In basic theory, CPM and PERT are largely the same. For both methods, the arrow diagram is the graphic model, and the mathematics are also quite similar. However, their important differences are as follows:

- CPM is used in situations where the duration times of activities are quite certain (e.g., in construction projects).

- PERT is used in situations where the duration times of activities are quite uncertain (e.g., in research and development projects).

For both of these methods to be most applicable, a project must possess the following four characteristics [28]:

- *Independent jobs or tasks.* More specifically, within a given sequence these jobs or tasks can be started, stopped, and performed independently.

- *Full job completion.* This means that whenever a job is started, it must continue without any interference until its total completion.

- *Ordered jobs.* This means jobs or tasks are ordered in a specified sequence.

- *The defined jobs' completion leads to actual project completion.* This means jobs or tasks are defined in such a manner that their completion leads to the end of the project.

The general steps associated with both PERT and CPM are shown in Figure 4.3, and the specific steps of each of these methods are presented next, separately.

4.7.1   PERT Steps

Usually, these seven steps are followed to develop a PERT network [28]:

1. Break down a given project into various jobs or tasks and identify each of these jobs or tasks.
2. Determine the sequence of these jobs or tasks and develop a network.
3. Estimate the duration time of each activity.
4. Obtain each activity's expected duration time.
5. Determine the time variance of each activity.
6. Determine the network's critical path.
7. Calculate the probability of project completion on a given time or date.

The following formulas are used in steps 4, 5, and 7, respectively [28, 29]:

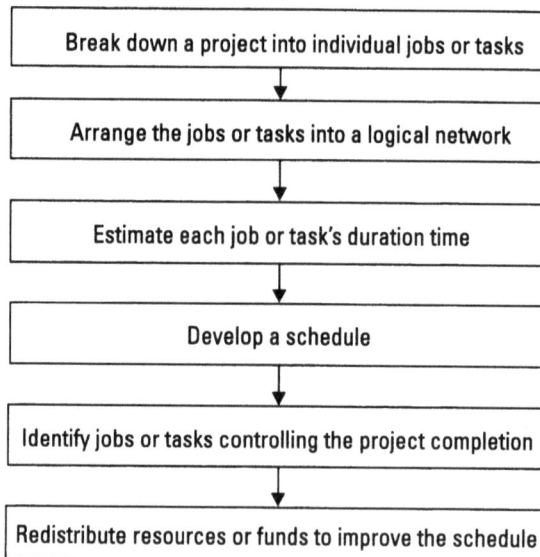

Figure 4.3 General steps of PERT and CPM.

Formula A

$$EADT = \frac{OADT + 4\ MLADT + PADT}{6} \tag{4.20}$$

where

    EADT   is the expected activity duration time.

    OADT   is the optimistic activity duration time. More specifically, this is the minimum time for completing an activity.

    PADT   is the pessimistic activity duration time. More specifically, this is the maximum time for completing an activity.

    MLADT is the most likely activity duration time. More specifically, this is the most likely time that it will take to complete an activity.

Example 4.5

Assume that the following data are given for a certain activity:

$$OADT = 30 \text{ hours}$$
$$PADT = 50 \text{ hours}$$
$$MLDT = 30 \text{ hours}$$

Calculate the expected duration time of the activity by using (4.20). By substituting the given data into (4.20), we get

$$EADT = \frac{30 + 4(35) + 50}{6}$$
$$= 36.67 \text{ hours}$$

The activity's expected duration time is 36.67 hours.

Formula B

$$\sigma_A^2 = \left[\frac{PADT - OADT}{6} \right]^2 \tag{4.21}$$

where σ_A^2 is the variance of the activity duration time.

Example 4.6

Using the given data in Example 4.5, calculate the variance of the activity duration time.

By inserting the specified data into (4.21), we get

$$\sigma^2_A = \left[\frac{50-30}{6}\right]^2$$
$$= 11.11$$

The variance of the activity duration time is 11.11.

Formula C

This formula is concerned with calculating the probability of completing a project on the due date or time. The following y transformation formula is used:

$$y = \frac{DT - EECT_\ell}{\sqrt{TS}} \qquad (4.22)$$

where

DT is the due time or date for completing the project.

$EECT_\ell$ is the earliest expected completion time of the last network activity.

TS is the total sum of variances of activity duration times along the network critical path (i.e., $TS = \sum \sigma^2_A$).

Equation (4.22) is associated with the normal distribution and Table 4.7 presents tabulated probability values for y.

Example 4.7

Assume that a PERT network representing a project is composed of 10 activities. The earliest expected completion time of the last network activity is 50 days and the sum of the variances of the activity duration times along the network critical path is 25 (i.e., $\sum \sigma^2_A = 25$). The project must be completed within 65 days. Calculate the probability that the project will be completed within the specified time.

By inserting the specified values into (4.22), we get

Table 4.7

Approximate Tabulated Values for the Cumulative Normal Distribution

Number	*y*	Probability
1	−4	0.00003
2	−3	0.00135
3	−2	0.02
4	−1	0.16
5	0	0.5
6	0.5	0.69
7	1	0.84
8	1.5	0.93
9	2	0.98
10	2.5	0.99
11	3	0.999
12	3.5	0.9998
13	4	0.9999

$$y = \frac{65 - 50}{\sqrt{25}}$$
$$= 3$$

From Table 4.7, for $y = 3$, the probability of completing the project within the specified time is 0.999.

4.7.2 CPM Steps

The following steps are associated with CPM:

1. Break down the project into various jobs or activities and identify them.
2. Determine the sequence of jobs or activities and develop the network.
3. Estimate duration time of each activity.
4. Determine the network's critical path.

4.7.3   CPM/PERT Symbols

The symbols used to construct a CPM or PERT network are shown in Figure 4.4. Each of these symbols is described next.

- *Circle.* This represents an event. More specifically, the circle denotes an unambiguous point in time in the life of a project (e.g., an event can be a start or a completion of an activity or activities). Normally, the events are labeled with a number.

- *Continuous arrow.* This denotes an activity that starts from a circle and ends at a circle. An activity for its completion requires money, time, and manpower.

- *Dotted arrow.* This is used to denote a restraint or a dummy activity. A dummy activity is an imaginary activity accomplished in zero time. Furthermore, the dummy activity does not consume any money, time, or manpower. Figure 4.5 shows an application of the dummy activity *y*. The figure denotes that activities L and M must be accomplished before starting activity N. However, only activity M must be accomplished prior to starting activity O.

- *Circle with divisions.* This also represents an event and is shown in Figure 4.4(d). In the figure, the circle is divided into two halves. The top half is used to label the event with a number, and the bottom half

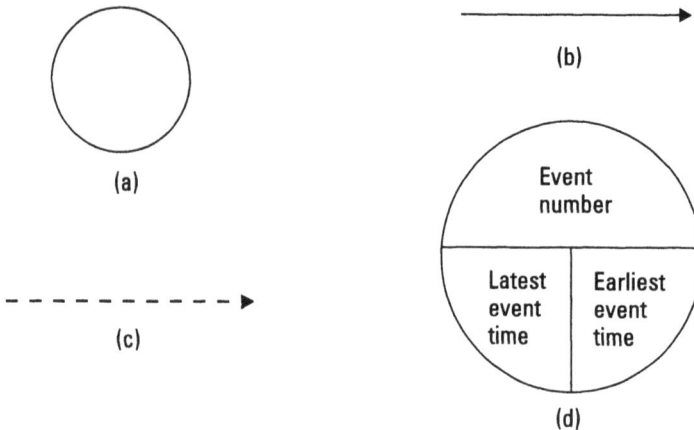

Figure 4.4 CPM/PERT symbols: (a) circle; (b) continuous arrow; (c) dotted arrow; and (d) circle with divisions.

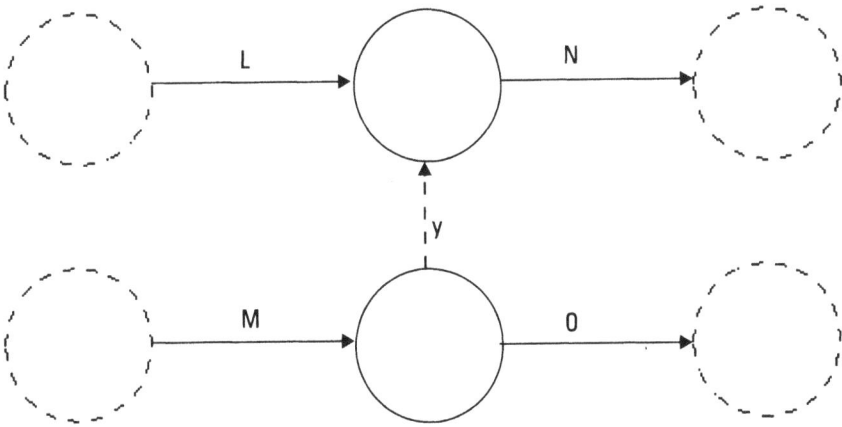

Figure 4.5 An example of the dummy activity application.

is divided further into two halves: the left side is used for the *latest event time* (LET) and the right side for the *earliest event time* (EET). EET means the earliest time in which an event can be reached, and LET the latest time in which an event can be reached without delaying the completion of the project.

4.7.4 Essential Formulas and Network Critical Path Determination

Various types of formulas are used in performing CPM/PERT network analysis. The symbology used in these formulas is shown in Figure 4.6.

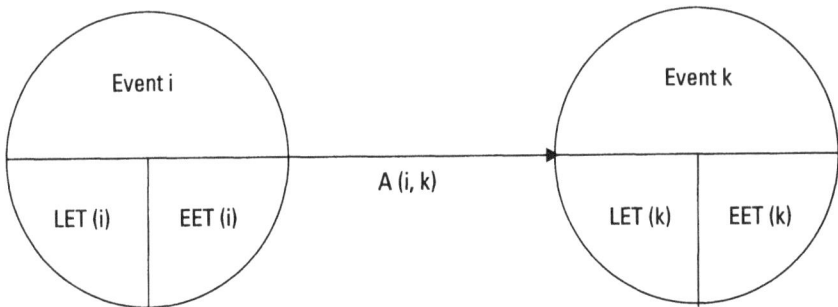

Figure 4.6 A network with a single activity.

The symbols in Figure 4.6 are defined here:

- $A(i, k)$ is the completion time of the activity between events i and k.
- $LET(i)$ is the latest event time of the event i.
- $LET(k)$ is the latest event time of the event k.
- $EET(i)$ is the earliest event time of the event i.
- $EET(k)$ is the earliest event time of the event k.

Formula I

This is used to calculate the latest start time for activity (i, k) expressed by

$$LST = LET(k) - A(i, k) \qquad (4.23)$$

where LST is the latest start time.

Formula II

This is used to calculate the latest finish time for activity (i, k) expressed by

$$LFT = LET(k) \qquad (4.24)$$

where LFT is the latest finish time.

Formula III

This is used to calculate the earliest start time for activity (i, k) expressed by

$$EST = EET(i) \qquad (4.25)$$

where EST is the earliest start time.

Formula IV

This is used to calculate the earliest finish time for activity (i, k) expressed by

$$EFT = EET(i) + A(i, k) \qquad (4.26)$$

where EFT is the earliest finish time.

Formula V

This is used to calculate the total float for activity (i, k) expressed by

$$TF = LET(k) - EET(i) - A(i,k) \qquad (4.27)$$

where *TF* is the total float.

The longest path from one end of a CPM/PERT network to another end (i.e., from the first event to the last event) is the critical path of the network. It means all activities along the critical path must be completed on time for project to finish on predicted time. It is to be noted that the term *longest path* means the total time of the critical path given by the largest sum of expected activity times of all paths that originate from the first event and terminate at the last event of the network.

A more systematic approach for determining the critical path of a CPM/PERT network is as follows:

- Construct network.
- Determine the EET of each network event by making a *forward pass* of the network and using the following relationship:

For any event k,

$$EET(k) = \text{maximum for all proceedings } i \text{ of } \left[EET(i) + A(i,k) \right] \quad (4.28)$$

The earliest event time of the first network event is equal to zero. More specifically,

$$EET(\text{first event}) = 0 \qquad (4.29)$$

- Determine the LET of each network event by making a *backward pass* of the network and using the following relationship:

For any event i,

$$LET(i) = \text{maximum for all proceedings } k \text{ of } \left[LET(k) - A(i,k) \right] \quad (4.30)$$

- The latest event time of the last network event and the earliest event time of the last network event are equal. More specifically,

$$LET(\text{last event}) = EET(\text{first event}) \qquad (4.31)$$

The latest event time of the first network event is always equal to zero [i.e., LET (first event) = 0].

- Identify the network events with $EET = LET$. If the network has only one path with events that satisfy the condition $LET = EET$, then that path is critical. Otherwise, go to the next step.

For each path whose events' $LET = EET$, determine the total floats for all of its activities using (4.27). Add the total floats of all activities of each path. The path resulting with the least sum of the total floats is the critical path.

Example 4.8

Assume that a project was broken down into seven activities, as presented in Table 4.8. The table also presents the expected duration times of all these activities. Construct a CPM network for the project and identify that network's critical path.

The CPM network, containing activity duration times, for the example is shown in Figure 4.7. The numerals in circles denote event numbers.

The network shown has the following three paths originating from event 1 and terminating at event 7:

- 1–3–5–6–7 (4 + 5 + 14 + 7 = 30 days);
- 1–2–3–5–6–7 (20 + 0 + 5·+ 14 + 7 = 46 days);
- 1–2–4–5–6–7 (20 + 8 + 6 + 14 + 7 = 55 days).

Table 4.8
Project Activities and Associated Data

Number	Activity	Immediate Predecessor Activity or Activities	Expected Duration Time (Days)
1	A	—	20
2	B	—	4
3	C	A, B	5
4	D	A	8
5	E	D	6
6	F	C, E	14
7	G	F	7

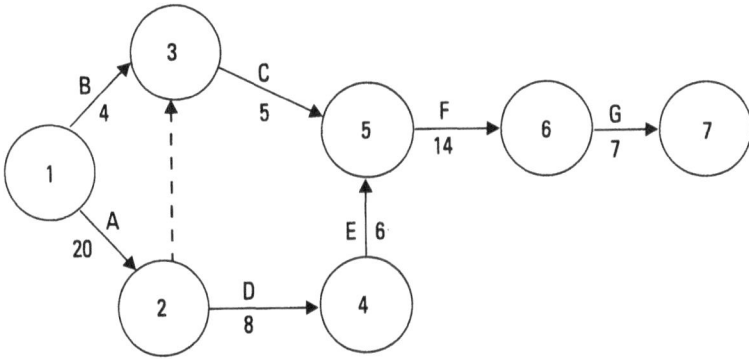

Figure 4.7 A CPM network for Example 4.8.

By examining the above paths, the last path will take the maximum time to complete the project. Thus, it is the longest and the critical path of the network (i.e., path 1–2–4–5–6–7). More specifically, the project can be accomplished in 55 days, if all activities along the critical path are completed within their specified times.

Example 4.9

Assume that the network diagram of a research and development project is shown in Figure 4.8. The diagram shows activities identified by letters, activity expected duration times (in days), earliest and latest event times for

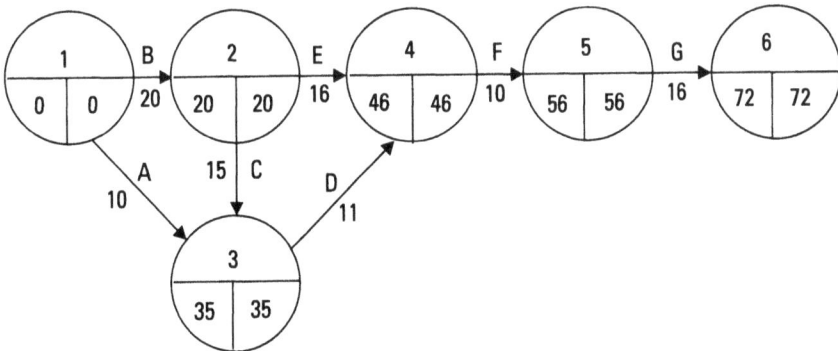

Figure 4.8 A network with given activity duration times along with determined earliest and latest event times.

each event, and event identification numbers. The optimistic, most likely, pessimistic, and expected time estimates for Figure 4.8 activities are given in Table 4.9.

Calculate the probability of completing the project in 77 days by using (4.21–4.22). The longest and the critical path of the network shown in Figure 4.8 is 1–2–3–4–5–6 (20 + 15 + 11 + 10 + 16 = 72 days). It means, the predicted time to complete the project is 72 days.

Table 4.10 presents the variance for each activity along the critical path. Its value was calculated using (4.21) and Table 4.9 data.

Table 4.9
Optimistic, Most Likely, Pessimistic, and Expected Time Estimates for Activities

Activity	Time Estimate for Each Activity (Days)			Activity Expected Time (Days) Using (4.20)
	Optimistic	Pessimistic	Most Likely	
A	8	16	9	10
B	15	25	20	20
C	11	25	14	15
D	8	18	10	11
E	10	26	15	16
F	6	18	9	10
G	8	28	15	16

Table 4.10
Variance for the Critical Path Activities

Activity	Variance δ^2_A
1–2 (B)	2.78
2–3 (C)	5.44
3–4 (D)	2.78
4–5 (F)	4
5–6 (G)	11.11
Total sum (TS) for all activities	26.11

By substituting the given data and the calculated data values of Figure 4.8 and Table 4.9 into (4.22), we get

$$y = \frac{77 - 72}{\sqrt{26.11}}$$
$$\simeq 1$$

Using the above calculated value and Table 4.7, we conclude that the probability of completing the project in 77 days is about 84%.

4.7.5   CPM Advantages and Disadvantages

Some of the advantages of the CPM are as follows [30]:

- It is useful to show interrelationship in work flow.
- It is useful in cost control and cost saving.
- It is useful to determine project duration systematically.
- It is useful to monitor project progress effectively.
- It is useful to avoid duplications and omissions.
- It is useful to improve communication and understanding.
- It is useful to determine the need for labor and resources in advance.
- It is useful to identify critical work activities for completing the project on time.

By contrast, some of the disadvantages of the CPM are high cost, time consumption, poor time-estimate provision, and a bias to use pessimistic time estimates.

4.8   Project Management Benefits, Obstacles to Achieve Benefits, and Project Management Failure Factors

There are many benefits of project management. Some of these are as follows [26]:

- It identifies a methodology for tradeoff analysis and time limits for scheduling.
- It minimizes the need for continuous reporting and improves the estimating capability for potential planning.

- It identifies functional responsibilities for ensuring that all activities are accounted for, irrespective of manpower turnover.
- It identifies problems early so that corrective measures may follow.
- It can measure accomplishments against plans.

The project management benefits can only be achieved by overcoming obstacles such as project complexity and risks, changes in technology, forward planning and pricing, special needs of customers, and organizational restructuring [26].

Past experience indicates that project management has failed because of factors such as [26]:

- Failure to inform employees about the workings of project management;
- Failure to explain the effect of the project management organizational structure on the wage/salary administration program;
- Failure to convince employees that executives completely support the change;
- Failure of executives to choose appropriate projects and project managers for the first few times and projects;
- The project management was not needed.

4.9 Problems

1. What are the advantages and disadvantages of the CPM?
2. What are the benefits of project management?
3. Define the following terms:
   - Project selection;
   - Project management;
   - Project.
4. Discuss the types of information required to evaluate a project.
5. What are the reasons for rarely using the mathematical models when selecting projects in the industrial sector?
6. Discuss three different types of project selection models.
7. What are the functions of project management?

8. What are the responsibilities of a project manager?

9. Write an essay on PERT and CPM.

10. A project was broken down into eight distinct activities as presented in Table 4.11. The table also presents the expected duration times of all these activities. Construct a CPM network for the project and identify the network's critical path.

Table 4.11
Project Activities and Related Data

Number	Activity	Immediate Predecessor Activity or Activities	Expected Duration Time (Days)
1	A	—	10
2	B	—	3
3	C	—	4
4	D	A, B, C	7
5	E	A	6
6	F	E	9
7	G	D, F	15
8	H	G	4

References

[1] Cleland, D. I., *Project Management*, New York: McGraw-Hill, 1994.

[2] Gaddis, P. O., "The Project Manager," *Harvard Business Review*, May–June 1959, pp. 55–60.

[3] Meredith, J. R., and S. J. Mantel, *Project Management*, New York: John Wiley and Sons, 1995.

[4] Wearne, S. H., *Control of Engineering Projects*, London: Edward Arnold Publishers, 1974.

[5] Banki, I. S., *Dictionary of Administration and Management*, Los Angeles, CA: Systems Research Institute, 1981.

[6] Thamhain, H. J., *Engineering Management*, New York: John Wiley and Sons, 1992.

[7] Brochure No. 7, Project Management Institute, Drexel Hill, PA.

[8] Liberatore, M. I., and G. I. Titus, "The Practice of Management Science in R and D Project Management," *Management Science,* August 1983, pp. 70–74.

[9] Gee, R. E., "A Survey of Current Project Selection Practices," *Research Management,* Vol. 14, No. 5, 1971, pp. 38–45.

[10] Thomas, H., "Some Evidence on the Accuracy of Forecasts in R and D Projects," *R and D Management,* Vol. 1, No. 2, 1971, pp. 55–69.

[11] Shannon, R. E., *Engineering Management,* New York: John Wiley and Sons, 1980.

[12] Souder, W. E., "Utility and Perceived Acceptability of R and D Project Selection Models," *Management Science,* August 1973, pp. 89–94.

[13] Dhillon, B. S., *Engineering Management,* Lancaster, PA: Technomic Publishing Company, 1987.

[14] Cetron, M. J., and J. Martino, "The Selection of R and D Program Content: Survey of Quantitative Methods," *IEEE Trans. on Engineering Management,* Vol. 14, 1967, pp. 4–13.

[15] Clarke, T. E., "Decision-Making in Technologically Based Organizations: A Literature Survey of Present Practice," *IEEE Trans. on Engineering Management,* Vol. 21, 1974, pp. 9–23.

[16] Baker, N. R., and W. J. Pound, "R and D Project Selection: Where We Stand," *IEEE Trans. on Engineering Management,* Vol. 11, 1964, pp. 124–134.

[17] Kasner, E., *Essentials of Engineering Economics,* New York: McGraw-Hill, 1979.

[18] Baker, N. R., "R and D Project Selection Models: An Assessment," *IEEE Trans. on Engineering Management,* November 1974, pp. 61–65.

[19] Souder, W. E., "Comparative Analysis of R and D Investment Models," *AIIE Transactions,* April 1972, pp. 20–26.

[20] Disman, S., "Selecting R and D Projects for Profit," *Chemical Engineering,* December 1962, pp. 87–90.

[21] Mottley, C. M., and R. D. Newton, "The Selection of Projects for Industrial Research," *Operations Research,* 1959, pp. 740–751.

[22] Murdick, R. G., and E. W. Karger, "The Shoestring Approach to Rating New Products," *Machine Design,* January 1973, pp. 86–89.

[23] Seiler, R. E., *Improving the Effectiveness of Research and Development,* New York: McGraw-Hill, 1975.

[24] Sobelman, S. A., *Modern Dynamic Approach to Product Development,* Dover, NJ: Picatinny Arsena, 1958.

[25] Martin, C. C., *Project Management: How to Make It Work,* New York: AMACOM, 1976.

[26] Kerzner, H., *Project Management: A Systems Approach to Planning, Scheduling, and Controlling*, New York: Van Nostrand Reinhold, 1992.

[27] Riggs, J. L., and M. S. Inoue, *Introduction to Operations Research and Management Science: A General Systems Approach*, New York: McGraw-Hill, 1975.

[28] Chase, R. B., and N. J. Aquilano, *Production and Operations Management: A Life Cycle Approach*, Chicago, IL: Irwin, 1981.

[29] Clark, C. E., "The PERT Model for the Distribution of an Activity Time," *Operations Research*, Vol. 10, 1962, pp. 405–406.

[30] Lomax, P. A., *Network Analysis: Application to the Building Industry*, London: The English Universities Press Limited, 1969.

5

Management of Engineering Design and Product Costing

5.1 Introduction

Humans have been designing engineering-related objects and structures for thousands of years, and two important examples of ancient civil engineering structures are the Egyptian pyramids and the Great Wall of China [1]. Since those days, we have progressed to the modern times of today in which each day hundreds of newly designed engineering products are put into service and old ones are discarded. Each new engineering product sold is designed within general guidelines such as meeting customer requirements and keeping the product's selling price within the buying power of the customers. It means the design of the product must be managed in such a manner that all the necessary design guidelines and other related factors are effectively satisfied.

Over the past 50 years, due to competitive environments, significant changes have occurred in the United States with respect to product cost. Thus, product costing has become very important to management because an engineering company will venture into developing a new product only if it will generate an acceptable level of profit. In order to estimate the acceptable level of profit, the cost of various aspects associated with the product under consideration must be estimated with care.

This chapter presents various important aspects of engineering design management and product costing.

5.2 Design Types and Approach

A design engineer may be expected to produce various different types of design. These may require varying degrees of ability, creativity, and management. Nonetheless, engineering designs may be grouped under the following three broad classifications [2, 3]:

- *Developmental design.* This type of design is initiated from the already existing design and may involve a considerable amount of technical work. The final designed product could be significantly different than the original one.

- *Creative design.* This is the design of a completely new item without any precedent whatsoever, and it requires a high degree of competence from the designers. A relative handful of designers will be employed in this type of design work.

- *Adaptive design.* This is concerned with adaptation of already existing designs to meet new challenges. Usually, most of the design work undertaken by engineers belongs to this classification. This type of design generally requires at least basic technical skills and some level of creativity, and usually the fresh designers go through the route of adaptive design work.

A general approach used in designing engineering products is mainly composed of the following eight steps [4]:

1. *Perform need analysis.* This first step of the design approach is concerned with analyzing the user needs (i.e., technical, physiological, psychological, social, and cultural). In order to produce an acceptable design from the user standpoint, needs such as these must be satisfied effectively.

2. *Define the problem.* This step is concerned with defining items such as objectives, boundaries, inputs and outputs, variables, and major restraints. More specifically, questions on areas such as those listed next are asked at this stage of the design:

   - Objectives to be satisfied by the design;
   - The largeness of the design problem;
   - The design problem under consideration;
   - Any particular difficulty;

- The anticipated social consequences;
- Items to be constraints in the problem definition;
- The design-associated constraints.

3. *Find effective alternative solutions to the problem.* This step is concerned with finding alternative solutions to the problem under consideration by taking advantage of past experience and creative methods such as those discussed in Chapter 7.

4. *Perform feasibility analysis of solutions.* This step is concerned with examining all alternative solutions with respect to their economic, physical, social, and cultural feasibility.

5. *Optimize most promising solutions.* This step is concerned with optimizing the most promising alternative design solutions. Approaches such as linear programming, a graphical model, calculus, and a physical model are used to optimize the engineering design.

6. *Choose the most promising solution.* This step is concerned with selecting the most promising solution by using the established design criteria for judging and evaluating the optimized alternative design solutions.

7. *Implement the most promising solution.* This is concerned with the implementation of the selected design solution through developing instructions, specification, and prototypes.

8. *Redesign.* This basically means to perform this task by following all of the above steps.

5.3 Expectations from a Design Department and Product Design–Related Areas Requiring Decisions

A company management has various expectations from its engineering design department, and the usual ones are shown in Figure 5.1 [5]. The expectation to "produce creative design" is concerned with developing designs that improve performance, reduce cost, and appeal to users. The expectation to "produce sellable design" means the design effectively satisfies technical, psychological, physiological, social, and cultural needs. The expectation to "create producible design" is concerned with ensuring that requirements such as available skilled manpower, technology, and facilities are carefully considered. The expectation to "produce reliable design" means that the design satisfies all specified reliability requirements, and it is very

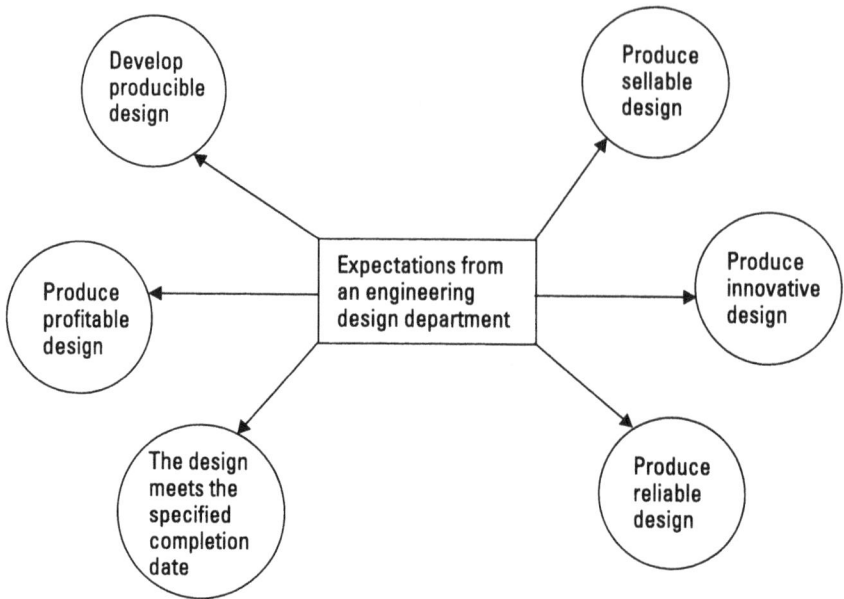

Figure 5.1 Management's expectations from its engineering design department.

reliable for its intended purpose. The expectation to "produce profitable design" is concerned with ensuring that the design will be profitable to the company by performing various types of financial analyses. The expectation that "the design meets the specified completion date" means that the design under consideration will be accomplished by the specified completion date without any delay whatsoever.

There are many product design–related areas that require decision. Some of these are as follows [6]:

- The type of design proposals required;
- The type and amount of documentation required;
- The responsibility for project;
- Breakdown of tasks or work;
- Responsibility for liaising with suppliers;
- Project priority;
- Evaluation of budget costs;

- The degree of responsibility for design appearance, production, test, and environment requirements;
- Division of responsibility among concerned technical areas (e.g., between electrical and mechanical design groups);
- The degree of research and investigation needed;
- The necessary approvals required.

5.4 Engineering Design Manpower and Tasks and Qualities of a Design Engineer

Depending on the type and size of an engineering product under design, the professionals involved, along with the design engineer, may vary from one design project to another. Table 5.1 presents examples of these professionals [1].

The safety engineer examines the design from the safety aspect by conducting various safety-related studies. The quality control engineer evaluates the design under consideration from the standpoint of quality, particularly when the designed product passes through the production phase. The reliability engineer examines the proposed design from the reliability aspect and ensures that the specified reliability parameters in the design specification document are fully satisfied. One example of these parameters is mean time between failures. The human factors engineer is concerned with ensuring that satisfactory consideration is given to the human element during the product design phase. The efforts of this individual during the design phase

Table 5.1
Examples of Professionals Who May Work Alongside a Design Engineer

Number	Professional
1	Safety engineer
2	Quality control engineer
3	Reliability engineer
4	Human factors engineer
5	Maintainability engineer
6	Manufacturing engineer

contribute directly to the effectiveness of the product's operation and maintenance.

The maintainability engineer examines the design from a maintainability angle. As the design specification of a product may specify the required mean time to repair, the maintainability engineer ensures this fulfillment by the final product. The manufacturing engineer is concerned with evaluating the design from the standpoint of determining its ease of manufacturability.

A design engineer performs various types of tasks. Some of these are as follows [1]:

- Design product.
- Optimize the design.
- Coordinate the design activity with all concerned individuals.
- Participate in design reviews.
- Produce new ideas for designs.
- Keep abreast of the changing environments and technology.
- Keep the design within given constraints.
- Record all changes.
- Keep management up to date with the design activity.
- Answer design-related questions.

In order to perform these tasks effectively, the designer must possess certain qualities. Some of these qualities are shown in Figure 5.2 [7, 8].

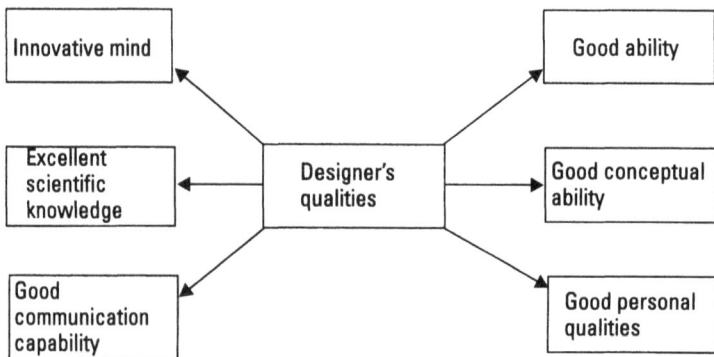

Figure 5.2 An engineering designer's qualities.

Good ability means the designer must be able to think logically during both the preparation of the overall design strategy and the actual design process. Good conceptual ability is concerned with the designer having the ability to comprehend and visualize design concepts. Excellent scientific knowledge means a good designer must possess superior scientific knowledge in the area of his or her specialization to produce good designs. Some of the basic areas in which the engineering designer should be well versed include technical drawings, chemistry, physics, economics, management, elementary and advanced mathematics, and engineering and technology. An innovative mind means the designer must possess a creative mind in order to address technical, economic, and managerial problems.

Good personal qualities include such items as concentration, capability, integrity, ability to tolerate criticism, perseverance, flexibility, willpower, good temperament, and good memory. Good communication capability is concerned with the designer being able to communicate effectively both verbally and in writing, thus the desired information can be disseminated to interested parties in an effective manner.

5.5 Design Reviews and Design Review Team, Team Chairperson, and Questions

During the product design phase or development cycle, various types of design reviews are conducted to ensure that the design effort is progressing satisfactorily with respect to design specification and plans. In general, design reviews may be grouped under the four classifications shown in Figure 5.3 [9].

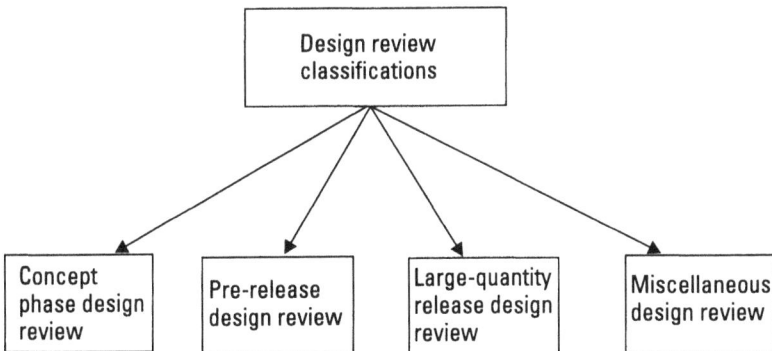

Figure 5.3 Design review classifications.

Nonetheless, a design review performed just prior to the release of engineering drawings is quite useful in many ways, including [10]:

- To assure that the product under consideration is designed to specifications;
- To determine the design from electrical, mechanical, and thermal aspects;
- To ensure that the interchangeability of like circuits, subsystems, and parts has been given adequate consideration during design;
- To identify causes that may lower product reliability;
- To ensure that the standard or preferred parts and preferred circuitry are used in an effective manner;
- To ensure that the effort put into the quality control area will turn out to be very effective;
- To ensure that the human factors have received adequate attention.

The main objective of the design review team is to examine item design from various aspects, and the team membership may be grouped under three person classifications: design specialists not directly involved with the product under consideration, technical personnel directly involved with the product under consideration, and representatives from outside agencies (if applicable).

As per [10], in order to obtain best results, the team membership should be limited to around 20 people. Nonetheless, usually the following technical specialists form the design review team [5, 11]:

- Engineer(s) involved in designing the product;
- Team chairperson;
- Design engineer(s) other than the one(s) whose work is under review;
- Manufacturing engineer(s);
- Tooling engineer(s);
- Reliability engineer(s);
- Quality control engineer(s);
- Field engineer(s);
- Material engineer(s);
- Packaging and shipping representative(s);

- Procurement representative(s);
- Customer representative(s).

The design review team chairperson chairs the design review team meetings. As this individual plays an important role during design reviews, a careful consideration must be given to such factors as technical competence, his or her position in the organization, and his or her personality. A design review chairperson performs tasks such as chairing the design review meetings, scheduling reviews, choosing items for design review, developing procedures for design reviews, deciding the nature or type of reviews, assisting and coordinating with bodies in preparation of necessary data, distributing review-related documents to all concerned people (e.g., drawings, agenda, and data files), distributing the minutes of the review meetings to all concerned individuals, and directing the appropriate follow-up actions after the review.

During reviews, the design is examined from various different perspectives. Past experience indicates that for a major product design, questions on areas shown in Figure 5.4 have proven to be very beneficial [9].

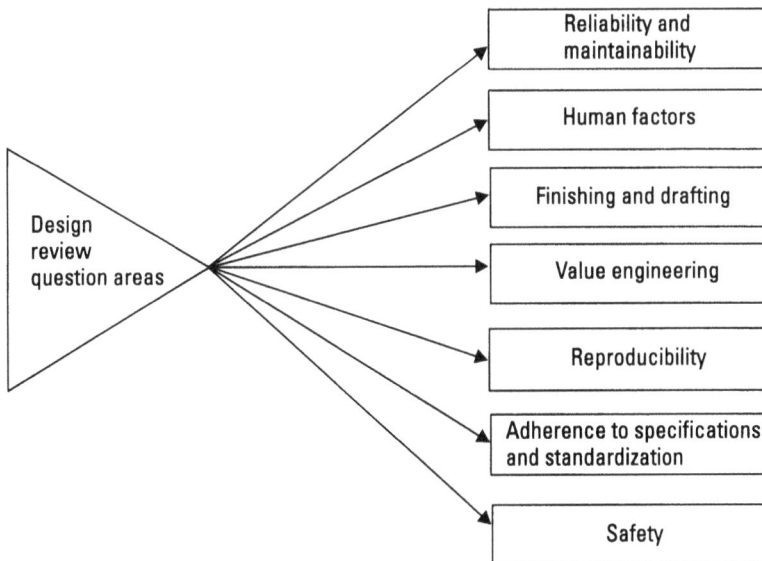

Figure 5.4 Important areas of design review questions.

Some examples of questions that can be asked on product reliability and maintainability are as follows:

- Under what assumptions was the product reliability estimated?

- Was the quantitative reliability specified in the design specification and is the design specification fully satisfied?

- Was the unit replacement maintenance considered?

- Will the product satisfy the specified value of downtime?

This is one example of the questions that can be asked with respect to human factors: Was appropriate consideration given to factors such as logical arrangement of controls, meters, and glare?

The following two questions are examples that can be asked on finishing and drafting:

- Was proper consideration given to factors such as paint finish and weld laps?

- Was proper consideration given to items such as dimension groupings, tolerances, and datum lines?

One example of the questions that can be asked on value engineering is: Is there any cheaper part that can perform the same function as the existing one?

An example of the questions that can be asked with regard to reproducibility is: Is it possible to reproduce the workable product model economically and efficiently in the production shop?

With regard to safety, an example of the questions that can be asked during a design review meeting is: Will the product be safe for humans?

Finally, two examples of questions concerning adherence to specifications and standardization are as follows:

- Is the right color recommended in finishing of the item under consideration?

- What was the reason to use nonstandard parts when the standard ones are readily available?

5.6 Engineering Drawing Types, Uses, and Guidelines for Drafting Managers When Producing Original Drawings with Computer-Aided Design (CAD)

In order to produce a final product, an engineering design must be translated into drawings. Thus, engineering design and drawings are closely interlinked.

Seven types of engineering drawings are frequently used in the industrial sector: field erection and assembly drawings, machining drawings, detail drawings, schematic diagrams, casting drawings, structural drawings, and installation drawings [11]. The field erection and assembly drawings are used to conduct assembly operations. The machining drawings are used to machine items to their specified requirements. The detail drawings are used to provide necessary information to produce and inspect a product. The schematic diagrams are used to depict characteristics, connections, and relationships of items under consideration. The casting drawings are used for casting in foundries. The structural drawings are used to provide information to manufacturing groups so they can develop well-processed specifications and standard times for manufacturing. The installation drawings are used to show, with respect to other parts, where and how an item is mounted.

Uses of engineering drawings include product design evaluation, maintenance, and modification; product manufacturing, testing, and installation; research and development documentation; component classification; product cost evaluation; the establishment of product selling price; the identification of product components; persuasion of management and customers, reliability evaluation, classification; maintenance manual preparation; and advertisement literature preparation.

Usually, drafting departments spend a significant amount of effort in producing design-related original drawings. The guidelines such as those listed next could be quite useful to drafting managers in producing original design-related drawings [12].

- Use functional drafting approaches.

- Become familiar with company's reproduction equipment;

- Become familiar with approaches used by the company to reproduce original drawings.

- Avoid drawing if a drawing can be traced.

- Avoid being overfunctional to the level that results in misleading drawings.

- Employ as much as possible the existing time-saving reproduction methods.
- Determine the fastest method to develop an original drawing under consideration.

Over the years CAD has emerged as an important tool in engineering and is being used successfully for various purposes. These include producing drawings, documenting designs, performing engineering analysis on geometric models, generating shaded images and animated displays, and carrying out process planning. Many organizations in the industrial sector have used CAD/*computer-aided manufacturing* (CAM) and have reported a variety of benefits. These include an improvement in productivity (depending on the nature of the task) from 25% to 350% with respect to using CAD systems against drafting boards, a reduction in development costs as high as 65%, and a reduction in the development cycle for new products (e.g., new cars from 5 years to 4 years) [1]. All in all, management in the area of engineering drawings can also take advantage of these new technologies to improve its effectiveness.

5.7 Reasons and a Procedure for Product Costing

In the design and manufacturing of products, companies must procure materials, employ manpower to perform work, pay for overhead costs, determine the capital requirements, and so on. All of these items directly or indirectly require cost estimations to make effective decisions. Thus, product costing is an essential element of design and manufacturing. Some of the specific reasons for product costing are as follows [13, 14]:

- To determine the most profitable material and procedure for manufacturing a product;
- To determine the capital required for equipment and other related items to manufacture a product;
- To verify bids from outside agencies;
- To conduct feasibility analyses of new products;
- To measure the efficiency of the product manufacturing process;
- To determine the cost-effectiveness of fabricating parts or assemblies or of procuring them from outside agencies;

- To establish product price;
- To determine whether or not it is economical to modify the existing facilities to manufacture the product under consideration;
- To determine the profitability of the product with respect to manufacture and market;
- To provide input to the long-term financial plans of the company;
- To assist in controlling the costs of product manufacturing.

Although no specific procedure can be used to obtain a reliable product cost estimate, a general approach that can usually be used to estimate a system cost is shown in Figure 5.5 [15].

The first step is "define problem" and is concerned with defining the product or system for which the cost is to be estimated. This is accomplished by establishing close links between cost and system analysts. The second step is "collect data" and is concerned with collecting various types of data that are important to estimate product cost. The third step is "estimate cost" and is concerned with performing actual cost estimate calculations. The fourth step is "present cost estimates" and is concerned with presenting the cost estimates of the preceding step in such a manner that they are easy and

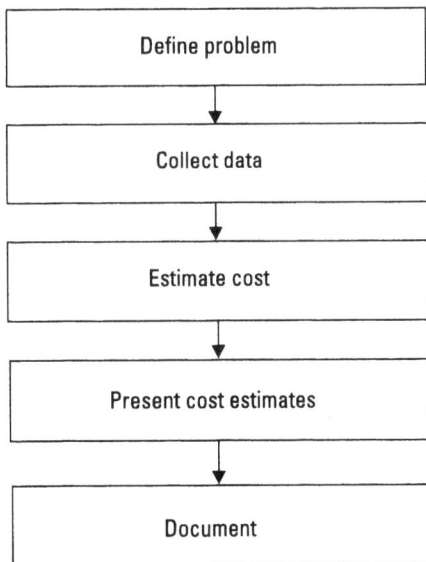

Figure 5.5 A general procedure for estimating system or product cost.

straightforward to use in making necessary decisions. The fifth step is "document" and is concerned with the documentation of the accomplished study.

5.8 Product Life Cycle Costing

Today the procurement decisions of many engineering products, particularly the expensive ones, are not entirely made on acquisition costs alone but on their life cycle costs. Past experience indicates that in many cases, the product ownership cost often exceeds the procurement cost. For example, [16] states that the product ownership cost (i.e., logistics and operating cost) can vary from 10 to 100 times the original acquisition cost. A high ownership cost can even be detected from the overall annual budgets of some organizations. For example, for 1974, the U.S. Department of Defense allocated 27% of its total budget to operation and maintenance, as opposed to 20% for procurement [17].

The life cycle cost may simply be defined as the sum of all costs incurred during an item's life span [18]. The history of the life cycle costing concept may be traced back to 1965, when the Logistics Management Institute prepared a report titled *Life Cycle Costing in Equipment Procurement* for the U.S. Department of Defense [19]. More specifically, the term *life cycle costing* was used for the first time in this document. A comprehensive list of publications on life cycle costing is available in [18].

Although there are various different general and specific models used to estimate life cycle cost, this section presents just two examples of these models: general and specific.

General Life Cycle Cost Estimation Model

In this case, the life cycle cost of an item is expressed by the following equation [18]:

$$LCC = RC_I + NRC_I \qquad (5.1)$$

where

LCC is the item life cycle cost.

RC_I is the item recurring cost. The major components of the recurring cost are labor cost, maintenance cost, inventory cost, support cost, and operating cost.

NRC_I is the item nonrecurring cost. The components of the nonrecurring cost are research and development cost, acquisition cost, LCC management cost, training cost, transportation cost, support cost, installation cost, reliability and maintainability improvement cost, qualification approval cost, and test equipment cost.

Example 5.1

A company using a system to manufacture certain engineering components is contemplating replacing it with a modern system. Three different systems are being considered for its replacement, and their corresponding data are presented in Table 5.2. Determine which of the three systems should be procured to replace the existing one with respect to their life cycle costs.

The expected failure costs of systems X, Y, and Z, respectively, are given by

$$FC_X = (0.06)(3,000)$$
$$= \$180$$

$$FC_Y = (0.07)(2,000)$$
$$= \$140$$

and

$$FC_Z = (0.03)(5,000)$$
$$= \$150$$

Table 5.2

Data for Three Systems Under Consideration for Procurement

Number	Description	System X	System Y	System Z
1	Procurement price	$100,000	$80,000	$120,000
2	Expected useful life in years	10	10	10
3	Annual operating cost	$4,000	$9,000	$2,000
4	Annual interest rate	10%	10%	10%
5	Annual failure rate	0.06	0.07	0.03
6	Cost of a failure	$3,000	$2,000	$5,000

Inserting the specified and calculated data into (3.13), we obtain

$$PVFC_X = 180\left[\frac{1-(1+0.1)^{-10}}{0.1}\right]$$

$$\approx \$1,106$$

where $PVFC_X$ is the present value of system X failure cost over its life span.

Similarly, the present values of machines Y and Z failure costs, respectively, over their useful lives are

$$PVFC_Y = 140\left[\frac{1-(1+0.1)^{-10}}{0.1}\right]$$

$$\approx \$860$$

and

$$PVFC_Z = 150\left[\frac{1-(1+0.1)^{-10}}{0.1}\right]$$

$$\approx \$922$$

The present values of systems X, Y, and Z operating costs over their useful life span, using (3.13) and the given data, are

$$PVOC_X = 4,000\left[\frac{1-(1+0.1)^{-10}}{0.1}\right]$$

$$\approx \$24,578$$

$$PVOC_Y = 9,000\left[\frac{1-(1+0.1)^{-10}}{0.1}\right]$$

$$\approx \$55,301$$

and

$$PVOC_Z = 2,000\left[\frac{1-(1+0.1)^{-10}}{0.1}\right]$$

$$\approx \$12,289$$

where

$PVOC_X$ is the present value of system X operating cost over its useful life.

$PVOC_Y$ is the present value of system Y operating cost over its useful life.

$PVOC_Z$ is the present value of system Z operating cost over its useful life.

The life cycle costs of systems X, Y, and Z, respectively, are

$$LCC_X = 100,000 + 1,106 + 24,578$$
$$= \$125,684$$

$$LCC_Y = 80,000 + 860 + 55,301$$
$$= \$136,161$$

and

$$LCC_Z = 120,000 + 922 + 12,289$$
$$= \$133,211$$

By examining the above three results, it is concluded that system X should be procured because its life cycle cost is the lowest.

Specific Life Cycle Cost Estimation Model

Obviously, this model is concerned with estimating the life cycle cost of a specific item. Thus, the life cycle cost of a switching power supply is expressed by [20]

$$LCC_{sp} = C_i + C_f \tag{5.2}$$

where

LCC_{sp} is the switching power supply life cycle cost.
C_i is the initial cost.
C_f is the failure cost.

The failure cost is defined by

$$C_f = \lambda T (RC + CS) \tag{5.3}$$

where

    λ is the unit constant failure rate.

    RC is the repair cost.

    CS is the cost of spares.

    T is the expected life of product.

The cost of spares is expressed by

$$CS = \alpha (USC) \tag{5.4}$$

where

    α is the fractional number of spares provided per active unit.

    USC is the unit spare cost.

5.8.1 Cost Estimation Models

In order to estimate various components of the recurring cost of an item, there are many mathematical models available in the published literature. This section presents three such models.

Model I

This model is concerned with estimating the labor cost of corrective maintenance. Thus, the labor cost of corrective maintenance is expressed by [18]:

$$LCCM = (CML)(SOH)\left[\frac{MTTR}{MTBF}\right] \tag{5.5}$$

where

    $LCCM$ is the labor cost of corrective maintenance.

    CML is the cost of maintenance labor per hour.

    $MTBF$ is the item's mean time between failures.

    $MTTR$ is the item's mean time to repair.

Example 5.2

Assume that a motor's mean time between failures and mean time to repair are 1,000 hours and 5 hours, respectively. The estimated annual operating hours of the motor are 6,000 hours. Calculate the annual corrective maintenance labor cost of the motor, if the hourly corrective maintenance labor rate is $30.

By substituting the given data into (5.5), we get

$$LCCM = (30)(6,000)\left[\frac{5}{1,000}\right]$$

$$= \$900$$

The annual corrective maintenance labor cost of the motor is $900.

Model II

This is a very useful model to approximate future cost estimates where cost data are available for similar equipment of a different capacity. The following relationship is used to estimate the cost of the desired item [18]:

$$C_d = C_s \left[\frac{CP_d}{CP_s}\right]^\alpha \qquad (5.6)$$

where

C_d is the cost estimate of the desired item.

CP_d is the capacity of the desired item.

C_s is the cost of the similar item, of capacity CP_s.

α is the cost capacity factor and its value varies for different items [21–23].

However, in the event of having no reliable data, past experience indicates that it is quite reasonable to assume the value of α to be 0.6.

Example 5.3

Assume that the annual maintenance cost of a 10-passenger vehicle is $1,000. Estimate the annual maintenance cost of a similar 20-passenger vehicle if the value of the cost capacity factor is 0.7.

Substituting the given data values into (5.6) yields

$$C_d = (1,000)\left[\frac{20}{10}\right]^{0.7}$$

$$= \$1,624.5$$

It means that the annual maintenance cost of the 20-passenger vehicle will be $1,624.50.

Model III

This model is used to estimate the operation cost of an *alternating current* (ac) motor. The cost to operate the motor is expressed by

$$ACTOM = \frac{(0.746)(HP_M)(AMOT)(EC)}{\sigma} \qquad (5.7)$$

where

$ACTOM$ is the annual cost to operate the motor.
HP_M is the motor horsepower.
σ is the motor efficiency.
$AMOT$ is the annual motor operational time expressed in hours.
EC is the electricity cost per kilowatt-hour ($/kWh).

Example 5.4

Assume a 30-horsepower ac motor is operated for 3,000 hours annually. The motor efficiency and the estimated cost of electricity are 80% and 3 cents per kilowatt-hour, respectively. Calculate the annual cost of operating the motor.
    By using the specified values in (5.7), we get

$$ACTOM = \frac{(0.746)(30)(3,000)(0.03)}{0.80}$$
$$= \$2,517.75$$

The annual cost of operating the motor is $2,517.75.

5.9   New Product Pricing

The appropriate pricing of new products is important because it influences such people as consumers, wholesalers, and public policy makers. In turn, the influence on these people usually affects factors such as revenue, profit, and unit sale. Usually, there are four specific factors that play an important role in pricing a product: cost, competition, demand, and market behavior. Nonetheless, the potential new product pricing objectives include increasing company credibility, creating interest in the new product, maximizing profit in the long run, maximizing profit in the short run, winning the consumers' confidence as being a "fair" producer, discouraging competitors from reducing their product prices, discouraging new entrants in the market, avoiding government interference, bringing the new product to a "visibility" level, and keeping middle persons' loyalty [24].

Over the years, various types of mathematical models have been developed to establish product price. One such model is presented next.

5.9.1 Product Pricing Model

This model is concerned with determining the optimal price of a product unit, and it makes use of price/demand relationship data [25, 26]. The model assumes that the demand for a product follows the linear relationship:

$$d = \theta + \lambda x \tag{5.8}$$

where

d	is the total demand for the product expressed as number of units.
x	is the price of a single product unit.
λ	is a parameter (i.e., the slope of the straight line).
θ	is a parameter, and its value is taken as the value of d when x is equal to zero.

The total cost of a product is expressed by

$$TC = FC + VC.d \tag{5.9}$$

where

TC	is the total cost of a product.
VC	is the product variable cost per unit.
FC	is the product fixed cost.

Thus, the total revenue from the product is expressed by

$$\begin{aligned} TR &= xd \\ &= x(\theta + \lambda x) \end{aligned} \tag{5.10}$$

where *TR* is the total revenue from the product.

By subtracting (5.9) from (5.10), we get the following expression for profit:

$$P = TR - TC$$
$$= [x(\theta + \lambda x)] - [FC + VC.d] \qquad (5.11)$$
$$= x\theta + \lambda x^2 - FC - VC.d$$

Inserting (5.8) into (5.11) yields

$$P = x\theta + \lambda x^2 - FC - (VC)(\theta + \lambda x)$$
$$= x(\theta - VC.\lambda) + \lambda x^2 - FC - (VC)\theta \qquad (5.12)$$

Differentiating (5.12) with respect to x yields

$$\frac{dP}{dx} = (\theta - VC.\lambda) + 2x\lambda \qquad (5.13)$$

Setting (5.13) equal to zero and then solving for x results in

$$x^* = \frac{VC}{2} - \frac{\theta}{2\lambda} \qquad (5.14)$$

where x^* is the optimal value of x. More specifically, (5.14) gives the optimal price of a product unit.

5.10   Problems

1. Discuss the following:
   - Creative design;
   - Adaptive design;
   - Developmental design.

2. What are management's expectations from its engineering design department?

3. What are the functions of a design engineer?

4. What are the qualities of a good engineering designer?

5. List technical specialists who form a typical design review team.

6. What are the important areas of design review questions?

7. What are the reasons for product costing?

8. Write an essay on life cycle costing and define the term *life cycle cost.*

9. List at least eight potential new product pricing objectives.

10. Assume that an electric transformer's mean time between failures and mean time to repair are 4,000 hours and 10 hours, respectively. The estimated annual operating hours of the transformer are 7,000 hours. Calculate the annual corrective maintenance labor cost of the transformer, if the hourly corrective maintenance labor rate is $25.

References

[1] Dhillon, B. S., Engineering Design: A Modern Approach, Chicago, IL: Irwin, 1996.

[2] Ray, M. S., *Elements of Engineering Design,* Englewood Cliffs, NJ: Prentice Hall, 1985.

[3] Matousek, R., *Engineering Design,* New York: John Wiley and Sons, 1963.

[4] Love, S. F., "Design Methodology," *Design Engineering,* April 1969, pp. 30–32.

[5] Dhillon, B. S., *Engineering Management,* Lancaster, PA: Technomic Publishing Company, 1987.

[6] Cain, W. D., *Engineering Product Design,* New York: John Wiley and Sons, 1969.

[7] Harrisberger, L., *Engineermanship: A Philosophy of Design,* Belmont, CA: Wadsworth Publishing Company, 1966.

[8] Dhillon, B. S., *Quality Control, Reliability, and Engineering Design,* New York: Marcel Dekker, 1985.

[9] Simonton, D. P., "Way a Design-Review Committee Pays Off Dividends," in *Management Guide for Engineers and Technical Administrators,* N. P. Chironis (ed.), New York: McGraw-Hill, 1969, pp. 110–115.

[10] *Engineering Design Handbook: Development Guide for Reliability, Part Two: Design for Reliability,* Washington, D.C.: Headquarters, U.S. Army Material Command, 1976.

[11] Rowbotham, G. E. (ed.), *Engineering and Industrial Graphics Handbook,* New York: McGraw-Hill, 1982.

[12] Irwin, W. J., "More Hints for the Drafting Manager," in *Management Guide for Engineers and Technical Administrators,* N. P. Chironis (ed.), New York: McGraw-Hill, 1969, pp. 150–152.

[13] Clugston, R., *Estimating Manufacturing Costs,* Boston, MA: Cahners Books, 1985.

[14] Doyle, L. E., "How to Estimate Costs of New Products," in *Management Guide for Engineers and Technical Administrators,* N.P. Chironis (ed.), New York: McGraw-Hill, 1969, pp. 40–43.

[15] Goldman, T. A. (ed.), *Cost-Effectiveness Analysis: New Approaches in Decision-Making,* New York: Frederick A. Praeger Publishers, 1971.

[16] Ryan, W. J., "Procurement Views of Life Cycle Costing," *Proc. Annual Symposium on Reliability,* 1968, pp. 164–168.

[17] Wienecke-Louis, E., and E. E. Feltus, *Predictive Operations and Maintenance Cost Model,* Report No. ADA078052, 1979, National Technical Information Service, Springfield, VA.

[18] Dhillon, B. S., *Life Cycle Costing: Techniques, Models, and Applications,* New York: Gordon and Breach Science Publishers, 1989.

[19] *Life Cycle Costing in Equipment Procurement,* Report No. LMI Task 4C-5, Logistics Management Institute (LMI), Washington, D.C., April 1965.

[20] Monteith, D., and B. Shaw, "Improved R, M, and LCC for Switching Power Supplies," *Proc. Annual Reliability and Maintainability Symposium,* 1979, pp. 262–265.

[21] Dieter, G. E., *Engineering Design,* New York: McGraw-Hill, 1983.

[22] Sheldon, M. R., *Life Cycle Costing: A Better Method of Government Procurement,* Boulder, CO: Westview Press, 1979.

[23] Desai, M. B., "Preliminary Cost Estimating of Process Plants," *Chemical Engineering,* July 1981, pp. 65–71.

[24] Oxenfeldt, A. R., "A Decision-Making Structure for Price Decisions," *Journal of Marketing,* Vol. 50, 1973, pp. 100–105.

[25] Hisrich, R. D., and M. P. Peters, *Marketing a New Product: Its Planning, Marketing, and Control,* Menlo Park, CA: Benjamin-Cummings Publishing Company, 1978.

[26] Harper, D. V., *Price, Policy, and Procedure,* New York: Harcourt, Brace & World, 1966.

6

Management of Proposals and Contracts

6.1 Introduction

Each day thousands of proposals are being written by engineering organizations to bring in new business with an ultimate aim of making profit. This means that proposals are one of the most crucial factors in the success or failure of engineering companies, and they must be prepared in such a way that they can win contracts for organizations at the most profitable terms and conditions. A proposal may simply be described as a document comprised of an offer from the bidder or contractor to the customer or owner to perform certain work by a stated time at a defined quality level for a certain sum of money by using its own facilities and manpower [1–3].

A contract may simply be called a legally enforceable promise, and in today's society contracts are an essential component to the conduct of business [4]. They provide information such as expectations from the parties signing the contract document, essential dates (e.g., project start and completion), and methods of payment for the services.

Usually, the law governing a contract is determined by the place where it was originally signed, unless it is otherwise clearly stated in the contract document. For a contract to be enforceable through the law, it must possess elements such as lawful subject matter, competent contracting parties, formulation as per the provisions of the law, and valid considerations given to all involved parties [5]. This chapter presents directly or indirectly the important management-related aspects of proposals and contracts.

6.2 Technical Proposal Types, Higher-Level Management Considerations in Proposal Development, and a Proposal Development Procedure

The many different types of technical proposals can easily be classified into the following three broad categories [6]:

- Unsolicited technical proposals;
- Solicited technical proposals;
- Technical brochures.

Unsolicited technical proposals are usually initiated by contractors without a formal *request for proposals* (RFPs) from customers. This type of proposal occurs in situations when contractors feel that there is a dire need for certain work or study and such work or study requires financing from an outside agency or body. Contractors may also learn informally that there is a definite need for submitting a proposal for a specific type of work. Usually research funds are secured from sources such as government organizations through unsolicited proposals.

The solicited technical proposals are submitted by the contractors to customers for work to be accomplished only after reviewing a request for a proposal from customers. The RFP contains information on work to be done, and it usually provides information on the presentation of the required information from the potential contractor.

The technical brochures may simply be called *general proposals*. They are used to describe a contractor's capabilities to prospective customers so they can take advantage of services offered by the contractor. Technical brochures are essentially promotional tools.

The upper management of the contractor considers various factors prior to preparing a proposal for new work. These factors include competition, profit, contract price, project-related risks, compatibility of the work involved in the proposal with company policy, legal complications, contract duration, resources required, effect on company image if the bid is unsuccessful, the project's importance to the company, proposal preparation cost, and the decision to bid low to win the contract [7].

Although the specific proposal development approach may vary from one company to another, the overall approach followed by these companies is basically the same. Thus, a general approach for writing engineering proposals is as follows [1, 8–10]:

- *Study proposal requests.* This step is concerned with reviewing the proposal request or the contents of a proposed unsolicited proposal to determine what is really required in the proposed project and to summarize it into a single sentence (if possible). The main purpose of this exercise is to make all concerned individuals clearly understand the proposal objective.

- *Prepare proposal outline.* This step is concerned with developing the proposal outline, and its basic purpose is to guide the thinking of persons involved in developing the proposal.

- *Distribute copies of the outline to all involved people.* This step is concerned with distributing the proposal outline to all concerned individuals for their comments and suggestions and then incorporating the useful feedback in the final outline document.

- *Assign duties for preparing the proposal document.* This step is concerned with assigning the proposal preparation duties to groups or individuals by carefully considering factors such as writing the summary and the problem requiring solution [1].

- *Review the written parts of the proposal as they are being completed.* This step is concerned with reviewing the written parts of the proposal as being completed with respect to factors such as clarity, technical accuracy, and sequence.

- *Type.* This step is concerned with typing the proposal. The step may appear to be simple and straightforward, but it requires a careful consideration to factors such as typographical errors and effective binding.

- *Review final draft.* This step is concerned with reviewing the final draft of the proposal with the aim of identifying any error or weakness.

- *Submit the final document.* This step is concerned with submitting the final proposal document to all concerned bodies or individuals.

6.3   Proposal Components and Format and RFP

Depending on the project under consideration, solicitation type, and customer, the bid proposals can vary from a simple business letter to multivolume documents comprised of hundreds of text pages, illustrations, and other supporting materials.

In general, the three main components of a typical proposal are as follows [11]:

- *Technical.* This component describes the procedure that the contractor will follow to meet the client's project requirements, including tools, methods and techniques, system to be employed, and project schedule.

- *Management.* This component describes the contractor with respect to factors such as organization, policies and procedures, background and qualifications, financial status, overall capabilities and experience, and staffing. The basic objective of this component is to show the competence of the contractor to perform the work satisfactorily.

- *Cost.* This component discusses the details used in estimating the work cost, and it provides sufficient backup to give the prospective client the confidence of its accuracy. It also may include information on past contractor cost performance.

Usually, for small projects, all of these components (i.e., technical, management, and cost) are combined under one cover. But, in the case of large projects or complex industrial products, each of these components can be organized into a separate volume. Usually the entire proposal is organized into two volumes: one for technical and the other for commercial or cost and management. Tables 6.1–6.3 present typical information included under technical, management, and cost sections, respectively, for a large project or complex industrial product [11].

Although the format of a proposal may vary from one organization or project to another, the basic format of a typical proposal remains essentially the same. In fact, it can simply be grouped into the following three areas [6].

1. Front matter format and contents:
    - Proposal title;
    - Notice of proprietary right;
    - Preface;

Table 6.1
Typical Information Included Under the Technical Section of
a Large Project or Complex Industrial Product Proposal

Item Number	Information Item
1	Introduction and background information on contractor or company
2	Professional manpower schedule
3	Details of contractor's financial control, engineering, and procurement departments
4	Experience list of similar items built by the contractor
5	Resumé of contractor's key personnel
6	Experience list of all items built by the contractor company
7	Project management philosophy
8	Pictures of items built by the contractor company
9	Draft contract

Table 6.2
Typical Information Included Under the Management Section of
a Large Project or Complex Industrial Product Proposal

Item Number	Information Item
1	Process description, schedule, and flow diagrams
2	Engineering and utility flow diagrams
3	Equipment list and data sheets
4	Services provided by contractor and client
5	Operating requirements
6	Contractor's or client's standards or specifications

- Table of contents;
- List of illustrations;
- List of symbols;
- Summary or abstract.

Table 6.3
Typical Information Included Under the Cost Section of
a Large Project or Complex Industrial Product Proposal

Item Number	Information Item
1	Price breakdowns (i.e., labor and materials)
2	Escalations (lump-sum contract), taxes, and royalty payment
3	Price for services offered
4	Payment schedule
5	Amount for work to be subcontracted
6	Optional equipment

2. Main content format:
   - Problem introduction;
   - Statement of work;
   - Organization of program;
   - Discussion;
   - Contractor and its manpower qualifications;
   - References or bibliography.

3. Back portion content format:
   - Appendix;
   - Resumé;
   - Proposal distribution list.

The RFP may simply be described as solicitation, and it may be directed to internal or external sources, a single vendor, or a general solicitation. Although the complexity of the project and various other factors may dictate the depth and amount of information contained in an RFP, usually an RFP provides essential information for potential bidders [12]. A typical RFP contains information on items such as [13]:

- Work statement;
- Applicable rules and regulations;
- Specifications including required quality to be satisfied;

- The point of contact;

- List of all deliverable items (e.g., hardware, software, and reports);

- Corporate certifications and representations;

- Required contents of technical, management, and cost proposals.

6.4   Classifications, Layout, and Benefits and Drawbacks of Engineering Specifications

Specifications are used when developing new engineering products. The size and other factors associated with these specifications may vary from one product to another. Although there are many different types of engineering specifications, they may be grouped under the following three broad categories [1]:

- *Category I.* The specifications belonging to this category are concerned with design and contain information on how the product requirements are to be satisfied.

- *Category II.* The specifications belonging to this category are called the construction specifications and they discuss requirements for areas such as legality, materials, and methods required in the actual manufacture of an item.

- *Category III.* The specifications belonging to this category are concerned with performance and are used to describe the resulting product requirements. However, these specifications do not present approaches for satisfying the requirements.

Although the specifications layout may vary for different situations, a general format is presented next [14].

- *Title page.* This includes information such as title, organization's name and address, and date.

- *Table of contents.*

- *Introduction.* This includes background information of a general nature on the specifications under consideration.

- *Scope.* This includes precise and accurate information on the product under consideration. Also, this section accurately outlines the requirement boundaries.

- *Affiliate documents.* This section presents other relevant documents concerning specifications under consideration.

- *Technical specification.* This section is the backbone of the overall specification document and contains information useful for effectively controlling areas such as construction, process, and performance. Usually, the requirements presented in the section are divided into subgroups such as design and construction, performance, materials, dimensions, acceptance conditions, spares, quality assurance, maintenance, reliability, and maintainability.

- *Other necessary information particulars.* This section presents information on documentary materials such as test reports, operation handbooks, and product maintenance drawings.

- *Appendix, miscellaneous, and index.* The specification document presents these three items separately.

As in the case of any other engineering item, the engineering specifications, too, have their benefits and drawbacks. Their benefits include aiding standardization, aiding better understanding of the exact needs of the customer, and aiding the production of better quality products [15]. By contrast, the engineering specifications' drawbacks are time consumption and possible restrictions on innovation.

6.5 Contract Provisions, Contract Type Determining Factors, and Contract Types

A contract has many important provisions. For example, as per the recommendations of the American Society of Civil Engineers, any engineering services contract should have provisions for the following items [5]:

- Date of signing the contract agreement;
- Details of obligations of the engineer to the client;
- Statement concerning start and completion dates of contract work;
- Compensation for the work to be carried out;
- Summary of the engineer's services scope;

- Names and addresses of the contracting parties and their associated descriptions;

- Payments for additional work;

- Copyrights/plans reuse;

- Cancellation provisions for services prior to the actual completion date;

- Compensation payment for service termination before the actual completion date.

Various types of contracts are signed to design, manufacture, or procure engineering products. The usage of a particular type of contract for a specific purpose depends on factors such as those presented in Table 6.4 [16].

The contracts used in the industrial sector may be classified into two broad categories: A and B. The category A contracts are grouped by method of determining contract cost, and the category B contracts are grouped other than by method of determining contract costs. The types of contracts belonging to category A are shown in Figure 6.1 [17].

Table 6.4
Contract Type Determining Factors

Number	Factor
1	Degree of competition
2	Urgency of project
3	Complexity of design involved
4	Past performance of contractor
5	Administrative costs involved
6	Degree of risk involved
7	Project's life span
8	Previous experience with the same contractor
9	Contractor's accounting procedure
10	The type of procurement under consideration
11	Inclination of contractor or customer toward a particular type of contract
12	Degree of difficulty faced in estimating procurement cost

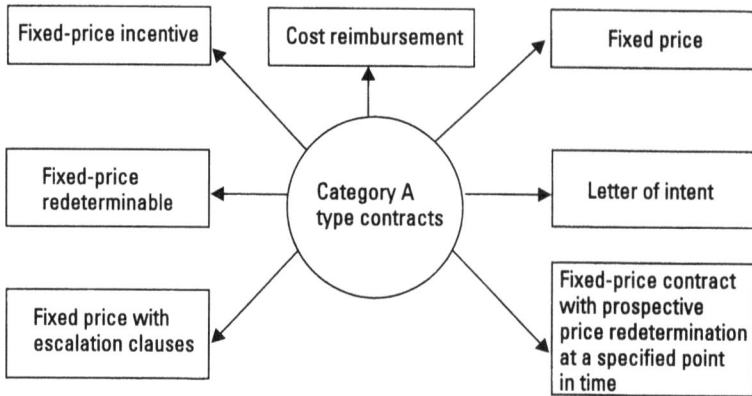

Figure 6.1 Category A type contracts.

The letter of intent is used as a preliminary contractual document prior to signing a formal contract. Its basic purpose is to authorize the contractor to start work on the project immediately, with the understanding that at a later stage a formal contract will be signed. Past experience indicates that usually a formal contract is signed within 60 days [1].

The fixed-price contract is used when a specified amount is paid in full after the successful completion of services or delivery of product to the customer. From the contractor's standpoint, this is probably the riskiest type of contract, because it has no provision to take into consideration increments in cost due to unforeseen events. Some of the benefits of the fixed-price contract are as follows:

- It is useful to provide maximum incentive to the contractor to reduce cost.

- It will increase contractor or manufacturer profit, if he or she can cut down the production cost.

- It provides maximum incentives to the contractor to develop material and labor usage saving procurement and production approaches.

All in all, the above advantages may have some overlap.

The fixed-price redeterminable contract is used in situations when accurate specifications and good estimates of labor and materials are not available. Thus, this type of contract is used to reduce the risk for a supplier or contractor risk. At some predetermined point(s) during the contract life, the contract price is adjusted either upward or downward.

The cost-reimbursement contract is used in situations when the performance cost cannot be estimated with certain accuracy. In this case, the customer pays the contractor the allowable contract costs. The cost-plus-fixed-fee and cost-plus-incentive-fee contracts are the two typical examples of a cost-reimbursement contract. One of the principal drawbacks of the cost-reimbursement contract is the administrative burden on the contractor and the customer.

The fixed-price contract with escalation clauses is used in situations when certain contingencies are outside the control of the supplier or contractor. Under such circumstances, various escalation clauses are added to the contract. These clauses make allowances for increases in the price of labor or materials in future.

The fixed-price contract with prospective price redetermination at a specified point in time is used in situations when only the initial phase of the project's cost is known. Subsequent to the initial phase, at a specified point during the remaining life of the contract, the price of the contract is redetermined, and it can either go upward to a specified ceiling or downward. This type of contract is often used in quantity production purchases.

The fixed-price incentive contract is used to provide incentives to contractors to reduce costs and improve efficiency without sacrificing product quality.

Some of the contracts belonging to Category B are as follows [18]:

- *Serial contract.* In this case, contractors and customers agree to undertake a series of separate contracts over a defined period.

- *Turnkey contract.* This is basically a package deal awarded to the main contractor, and in turn the main contractor may subcontract some specialized work to suitable subcontractors. The main advantage of this type of contract from the customer's standpoint is that the customer concerned is relieved from coordinating the works of subcontractors and other related details.

- *Competitive contract.* This is negotiated through the formal competitive tendering. More specifically, in this case a number of tenderers compete for the same project, and the customer awards the contract to the most suitable tenderer.

- *Package contract.* This is also known as the package deal, and in this case various related projects are combined and then awarded as a package deal.

- *Continuation contract.* This is used when a contractor is doing a good job on an ongoing project and, on that basis, another project is negotiated with the same contractor subject to same terms and conditions as those of the ongoing project.

- *Negotiated contract.* This is an alternative to the competitive contract, and in this case the customer negotiates the contract with a contractor of his or her choice. Usually, this type of contract is used in situations when the specifications of the project to be contracted are rather unclear.

6.6 Engineering Contract Documents and Contractor Selection Factors

In reality, there are many documents associated with a contract. A well-written contract usually clearly states the items to be regarded as the contract documents. These items include the following [18]:

- *Specification.* This describes items such as work description, design-related information, performance standards to be satisfied, maintenance requirements, and test procedures.

- *The tender.* This is basically the contractor's "offer" for eventual acceptance by the customer without any conditions whatsoever. However, the tender specifies conditions to which the offer is subjected, as well as items such as the payment method and the terms and price of the contract.

- *Formal agreement.* This is the confirmation of a contract in an official agreement and contains information on items such as payment terms, contract price, contract start and completion dates, and contract document listing.

- *Contract statement scope.* This is concerned with the contract subject matter and describes the customer requirements in a broad perspective. At the time of inviting tender, the customer makes such statements.

- *Contract conditions.* These are the rules by which the contract is governed, and they vary from one contract to another.

- *Bills of quantities.* These are issued by the customer usually when inviting tender, and one example of their usage is to itemize services for which the tenderers must provide prices.

- *Work execution–related data.* These data affect work execution and some of their examples are site conditions, geological data, and constraints on work hours.
- *Work schedule.* This is concerned with recording the critical dates associated with the contract during its lifespan. The work schedule is useful in monitoring the contractor's progress.
- *Work site rules.* These rules are imposed by the customer on the contractor for observing at the work site, and they may concern areas such as labor relations, noise, site cleanliness, and fire safety.

In order to select a right contractor for a job, the customer examines various factors. Table 6.5 lists most of these factors.

6.7 Contract Negotiation Approach and Negotiator's Characteristics

The task of negotiating a contract is very important for the successful completion of a project. Thus, it must be accomplished systematically. The steps of a contract negotiation approach are as follows [3]:

Table 6.5
Factors Examined During Contractor Selection

Number	Factor
1	Work capacity of the contractor
2	Quality of the contractor's manpower
3	Performance reputation of the contractor
4	Experience of contractor's manpower
5	Current unfinished workload of the contractor
6	Annual average volume of work of the contractor
7	Part and equipment availability of the contractor
8	Contractor's working capital
9	Contractor's past experience with the specific project
10	Fairness reputation of the contractor
11	Past record of the contractor in managing subcontractor's work

- *Solicit proposals.* This is concerned with soliciting proposals from desirable contractors, and its main objective is to obtain as many bids as possible so that the customer can choose the most desirable contractor at the most favorable terms and work conditions.

- *Evaluate proposals.* This is concerned with evaluating the submitted proposals from various different aspects and comparing the total ratings of all proposals to provide input to the proposal selection decision.

- *Choose contractor(s).* This is concerned with selecting the most desirable contractor with the aid of factors such as previous experience, knowledge of contractor's organization, and the performance reputation of the contractor.

- *Bargain for most desirable terms and work conditions.* This is probably the most important step of the contract negotiation approach and is concerned with bargaining or negotiating for the best terms and conditions from the customer standpoint. An able negotiator or bargainer plays a crucial role in this step.

- *Request higher management for contract.* After settling all terms and conditions with the most desirable contractor, this next step is concerned with requesting higher management for the actual contract document.

- *Sign contract.* This is concerned with awarding the contract to the most desirable contractor, and appropriate customer and contractor officers sign the final contract document.

As a contract negotiator is instrumental in the success or failure of a contract negotiation, careful consideration must be given when selecting an individual for this purpose. A good contract negotiator should possess characteristics such as those shown in Figure 6.2 [19].

6.8   Mathematical Models for Contracting

In contracting, various types of mathematical models are used to make sound decisions. This section presents two examples of such models.

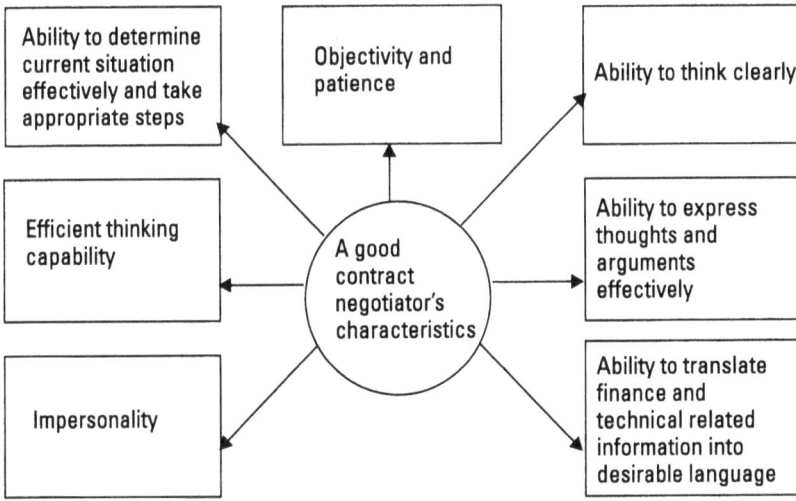

Figure 6.2 Qualities of a good contract negotiator.

Model I

This is a useful model to determine the price adjustment factor for labor. The price adjustment factor for labor is expressed by [20]

$$PAFL = \theta_I \left[\frac{LR_I - BLR_I}{BLR_I} \right] + \theta_{II} \left[\frac{LR_{II} - BLR_{II}}{BLR_{II}} \right] \qquad (6.1)$$

where

PAFL	is the price adjustment factor for labor.
LR_I	is the labor force type I labor rate for the month under consideration.
BLR_I	is the labor force type I base labor rate.
θ_I	is the labor force type I weighting factor (this type of labor force may be representing a specific trade).
θ_{II}	is the labor force type II weighting factor (this type of labor force may be representing general labor force).
LR_{II}	is the labor force type II labor rate for the month under consideration.
BLR_{II}	is the labor force type II base labor rate.

Example 6.1

A company categorizes its labor force into two groups: I and II. For each of these groups, we have the following data:

$$\theta_I = 0.5, LR_I = 113, BLR_I = 100,$$
$$\theta_{II} = 0.3, LR_{II} = 65, BLR_{II} = 50.$$

Calculate the price adjustment factor for labor. Substituting the above values into (6.1) yields

$$PAFL = 0.5\left[\frac{113-10}{100}\right] + 0.3\left[\frac{65-50}{50}\right]$$
$$= 0.155$$

It means the value of the price adjustment factor for labor is 0.155.

Model II

This is another useful model used to determine increase or decrease in contract price because of the changes in material and labor costs. The following equation is used to determine increase or decrease in *contract price* (CP) [18, 20]:

$$CP = OCP\left[\frac{(FCPM)(MIPA)}{(MIBD)} + \frac{(FCPL)(LIPA)}{(LIBD)} + NVCP\right] - OCP \quad (6.2)$$

where

MIPA	is the material index applicable at the time of price adjustment.
MIBD	is the material index at the base date. More specifically, this value is used by the contractors to determine the price given in the tender.
FCPM	is the fraction of the contract price that varies with materials.
LIPA	is the labor index applicable at the time of price adjustment.
LIBD	is the labor index at the base date. More specifically, this value is used by the contractors to determine the price given in the tender.

FCPL   is the fraction of the contract price that varies with labor.

OCP   is the original contract price.

NVCP   is the nonvariable part of the contract price. More specifically, this is given as a fraction of the contract price.

6.9 Problems

1. Define the following terms:

   - Proposal;

   - Contract.

2. Discuss the types of technical proposals.

3. Describe an approach used for writing engineering proposals.

4. What are the principal components of a proposal?

5. Describe a request for proposal.

6. Discuss three main classifications of engineering specifications.

7. What are the advantages and disadvantages of engineering specifications?

8. What are the factors useful in determining the type of contract to be pursued?

9. What are the principal factors used in selecting a contractor?

10. What are the important characteristics of a good contract negotiator?

References

[1]   Hicks, T. G., *Successful Engineering Management,* New York: McGraw-Hill, 1962.

[2]   Dunham, C. W., *Contracts, Specifications and Law for Engineers,* New York: McGraw-Hill, 1971.

[3]   Dhillon, B. S., *Engineering Management: Concepts, Procedures, and Models,* Lancaster, PA: Technomic Publishing Company, 1987.

[4]   Bagley, C. E., et al., "Legal Issues," in *Technology Management Handbook,* Boca Raton, FL: CRC Press, 1999, pp. 10.1–10.6.

[5]   Abbett, R. W., *Engineering Contracts and Specifications,* New York: John Wiley and Sons, 1963.

[6]   Reisman, S. J. (ed.), *A Style Manual for Technical Writers and Editors,* New York: Macmillan, 1962.

[7] Walton, T. F., *Technical Manual Writing and Administration,* New York: McGraw-Hill, 1968.

[8] Asner, M., "Management: Winning Proposals," *Computing Canada,* Vol. 15, July 6, 1989, pp. 18–20.

[9] Barakat, R. A., "Story Boarding Can Help Your Proposal," *IEEE Trans. on Professional Communications,* Vol. 32, March 1989, pp. 20–25.

[10] Asner, M., "Creating a Winning Proposal," *Business Quarterly,* Vol. 55, No. 3, 1991, pp. 36–40.

[11] Thamhain, H. J., *Engineering Management,* New York: John Wiley and Sons, 1992.

[12] Roman, D. D., *Managing Projects,* New York: Elsevier Science Publishing Company, 1986.

[13] Kezsbom, D. S., D. L. Schilling, and K. A. Edward, *Dynamic Project Management,* New York: John Wiley and Sons, 1989.

[14] *A Guide to the Preparation of Engineering Specifications,* The Design Council, London, 1980.

[15] Whittemore, H. L., "Ideas on Specifications," *Columbia Graphs,* 1952.

[16] Hajek, V. G., *Management of Engineering Projects,* New York: McGraw-Hill, 1977.

[17] Hayes, G. E., and H. G. Romig, *Modern Quality Control,* Encino, CA: Bruce: A Division of Benziger Bruce and Glencoe, Inc., 1977.

[18] Horgan, M. O. C., and F. R. Roulston, *The Elements of Engineering Contracts,* Report by W. S. Atkins and Partners, Woodcote Grove, Epsom, Surrey, U.K., August 1977.

[19] Dhillon, B. S., *System Reliability, Maintainability, and Management,* Princeton, NJ: Petrocelli Books, Inc., 1983.

[20] Marsh, P. D. V., *Contracting for Engineering and Construction Projects,* Aldershot, England: Gower Publishing Company Limited, 1981.

7

Creativity and Innovation

7.1 Introduction

The development of our modern civilization to its current level is the result of creative thinking of many individuals of past and present times. Some of the important examples of past inventions and discoveries are the wheel, the automobile, the telephone, the steam engine, radio and television, the airplane, and the Internet. Past experience indicates that important discoveries and inventions are generally not accidents, as they may appear to be, but are the result of the intense efforts and determination of intelligent individuals, which were overlooked by ordinary minds.

In today's competitive environment, the survival of engineering and technology organizations very much depends on creative thinking and innovation. Furthermore, some companies estimate that as much as 80% of their total sales is due to products that were not on the market a decade earlier [1]. This clearly shows the importance of creativity and innovation to the health of organizations.

Over the centuries, many publications on creativity have appeared, and [2] lists 6,823 references appearing during the period 1566–1974 alone. These references are grouped into many distinct areas: scientific creativity, general creativity, creativity in the fine arts, creativity of women, creativity and psychopathology, facilitating creativity through education, and creativity in industry, engineering, business, and developmental studies.

Current information on the history of research of creativity and the last 50 years of creativity research is given in [3, 4], respectively. This chapter presents important aspects of creativity and innovation considered useful to technical professionals.

7.2 Creativity and Invention Definitions, Classifications of Inventions, and Factors in Creativity

Both creativity and invention may simply be defined as follows [5, 6]:

- *Creativity.* This is the ability to produce interesting, useful, and new results.
- *Invention.* This is something novel and useful.

Over the years, many people have studied inventions and grouped them under seven distinct classifications [5, 7]:

1. *The simple or multiple combination of existing inventions.* This category contains inventions that are the result of a simple combination of two already existing inventions used to produce a new or better result.

2. *Application of a new principle to an old problem.* This category contains inventions that are the result of the application of a new principle to an old problem. More specifically, an old problem was solved with an available principle of the time and the application of a new principle achieved startling results. For example, the miniaturization of electronic components created a revolution in various areas of technology.

3. *Labor-saving concept inventions.* This category contains inventions in which an existing process/mechanism is altered to save effort, dispense with an operator, or produce more with the same effort.

4. *Adaptation of an old existing principle to an old problem to achieve brand new results.* This category contains inventions in which the problem has been in existence for a while, and the scientific principle key to its solution also has been known.

5. *Application of a new principle to a new use.* This category contains inventions in which new principles are applied to totally different areas of technology.

6. *Direct solution to a problem.* This category contains inventions in which a need was confronted by deliberately designing an item that satisfied the need in question.

7. *Serendipity inventions.* This category contains inventions that are results of accidental findings.

People may question the effects of factors such as age, sex, and education on creativity [8]. Here, we briefly examine each of these factors.

With respect to age, Alex Osborn stated, by citing various examples, that older persons are no less imaginative than younger ones [8]. In fact, he quoted some experts who said, "Creative imagination is ageless" and "Imagination grows by exercise." However, on the other hand, as per [9], the best ages for basic innovations among various specialists vary (e.g., experimental physicists: 35–40 years, medical and biological scientists: 40–45 years, and mathematicians and theoretical physicists: 30–35 years).

With respect to sex, it may be said that the women may be inferior to men in muscle mass, but not in imagination. In fact, various scientific studies indicate that women on average have faired better in imagination than the men [8].

With respect to education, various studies on creative aptitude clearly indicate that there is little or no difference, among individuals of the same age group, between those having college-level education and those without [8]. Furthermore, by examining the past inventions and discoveries, it may be said that many important ideas came from individuals who were devoid of specialized training in the problem concerned. Two typical examples are the invention of the telegraph by Morse, a professional portrait painter, and the invention of a new shell-fragment detector by an unscientific person [8].

7.3   Creativity Climate, Ways and Guidelines to Develop Creativity, and Creative Problem-Solving Processes

Climate is an important factor in the creativity of individuals. Obviously, better climate leads to better creativity results in an organization. Management can take a number of steps to create positive environments for creativity in a department. Some of these steps are as follows [10]:

- Provide appropriate facilities to highly creative individuals.
- Provide appropriate opportunity and freedom to deserving individuals.

- Do not resist new ideas.

- Reward highly creative individuals through financial means.

- Send highly creative individuals to professional conferences and seminars so they can exchange thoughts with their outside peers.

- Reward regularly creative individuals with status.

- Recognize variations in personality.

- Hold regularly informal seminars.

- Assign problems to individuals to be solved only after considering their best field of interest.

- Give recognition to new ideas as soon as possible and keep idea originator(s) current with related decisions.

Over the years, the creativity professionals have proposed various ways to develop creative talent. Six of these are shown in Figure 7.1 [8, 11].

Experience is the richest fuel for creativity, and it can be categorized into two main areas: firsthand and secondhand. Firsthand experience provides a richer fuel for creativity than the secondhand type. One gains secondhand experience through listening and superficial reading.

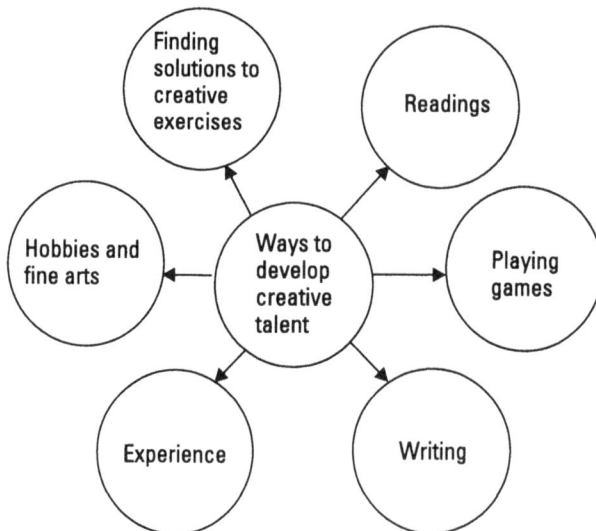

Figure 7.1 Ways to develop creative talent.

Writing is a very useful exercise to train the imagination or develop creativity. Various scientific tests rate "facility in writing" as a fundamental index of creative aptitude. Alex Osborn, in [8], argued the usefulness of writing in developing one's creative talent by citing many real life examples.

The right types of readings are rich in creativity-related vitamins. Magazines such as *Reader's Digest, National Geographic,* and *Holiday* are useful to fill the fuel tank of one's imagination.

Playing appropriate games effectively can help to develop one's imagination. However, it should be noted that there are approximately 250 types of sedentary games and only about 50 entail creative exercises [8].

Finding solutions to creative exercises is the most direct way to develop one's creative talent. To develop creativity talent, practice creativity. Some studies indicate that the individuals who have taken courses in creative problem solving were far better in the production of good ideas than the ones without the benefits of such courses.

Hobbies and fine arts are also very useful to develop creative talents. There are around 400 different types of hobbies, but most of them have more to do with collecting than creating. Collecting hobbies strengthens knowledge and trains judgment rather than stimulates the imagination. By and large, handicraft-related hobbies provide creative exercises to a greater degree than collecting.

Over the years, many professionals have proposed various guidelines for improving the creative thinking of an individual [5, 12–14]. Some of the important ones are presented here [15].

- Develop a positive attitude toward creative thinking.

- Unlatch inherent imagination by using methods available in literature, such as asking "what if" and "why."

- Develop persistence to the highest degree. Remember, Thomas Edison tested over 6,000 materials before discovering a bamboo species as a filament for the incandescent lightbulb—subsequently, Edison remarked, "An invention is 5% inspiration and 95% perspiration."

- Develop a supporting attitude toward a new idea, regardless of its origin.

- Avoid passing critical judgment on an emerging idea.

- Establish boundaries for the problem under consideration.

The creative problem-solving process is composed of six basic steps, as shown in Figure 7.2 [8, 11, 16]. Problem definition is concerned with the identification of the creative problem, and the preparation step deals with the collection and analysis of desirable data. The idea generation step calls for the production of many ideas, and idea development is concerned with selecting the most appropriate ideas. The evaluation step calls for the verification of tentative solutions, and the implementation step is the final solution adoption.

7.4 Types of Barriers to Creative Thinking, Management Barriers to Creativity, and Innovation Prevention Reasons

There are many barriers that may inhibit creativeness in any individual. The effective results can only be obtained after removing these barriers altogether. For practical purposes, these barriers can be grouped into four main classifications as follows [10, 11, 17, 18]:

- Perceptual barriers;
- Environmental barriers;
- Emotional barriers;
- Cultural barriers.

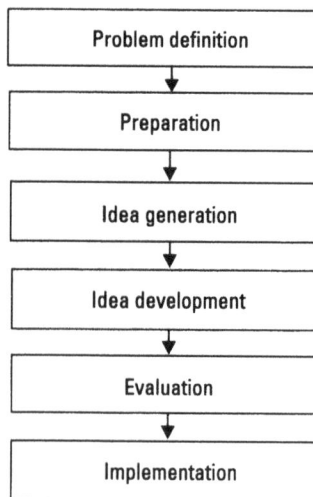

Figure 7.2 Creative problem-solving steps.

The perceptual barriers are the result of the mind's tendency to short circuit, and they occur because of failure to recognize each situation's components as individual elements. A typical example is that when an individual becomes overacquainted with some item, then he or she tends to overlook such an item's background in detail.

The environmental barriers are concerned with the development of an individual. A typical example of such a situation is when a person is fed with solutions to his or her problems regularly, say, by his or her parents, then that person generally becomes dependent upon others to find answers to his or her problems. The same reasoning may also be extended to cases such as boss-subordinate and student-teacher. The emotional barriers are probably the most serious barriers of all, and some of their examples are greed, fear, hate, poor self-confidence, desire for security, overcooperation, unwillingness to accept help from others, love, resistance to change, and self-satisfaction. The cultural barriers are the ones imposed by society on an individual. Their examples include conformity, competitiveness, and cooperation.

Management plays a key role in the enhancement of creativity. Although some of its actions could be necessary and desirable, others may very well be detrimental to creativity. Undesirable management barriers include the following [6]:

- Tendency to tell professionals or others rigidly what to do and how to do it;
- Practice of making frequent changes to key decisions;
- Reluctance to take chances;
- Too rigid a line of authority;
- Expression of a negative attitude towards all new ideas;
- Failure to recognize and reward creative individuals;
- Poor communication between management and workers;
- Nonexistence of long-range objectives;
- Improper handling of credit for new ideas;
- Placing of too much emphasis on quick or immediate utility of ideas.

In the case of mass-produced products, there are numerous reasons for the delay or prevention of innovations, including [19]:

- Consumer resistance;
- Unavailability of capital;

- Inadequate competition;
- Sociological implications (e.g., codes for safety and overhauling regulatory laws);
- Current commitments of the organization in question.

7.5 Individual Creative Person, Engineer, and Manager Characteristics; Attributes of a Manager of Creative People; and a Noncreative Person's Characteristics

Over the years many studies have been performed to determine the typical characteristics of a creative person. As per these studies, the typical creative person works relentlessly, possesses good sense of humor, is always open to experience, possesses listening ability, accepts failures easily, places no value on job security, is a nonconformist and accepts chaos, is independent and is insensitive to the feelings of others, possesses an IQ between 100 and 140, is observant and takes interest in exploring ideas, likes to seek privacy and autonomy, and does not give any importance to status symbols [20–22].

Similarly, typical highly creative engineers possess observation and concentration power, are independent in thought and action, are less anxious, belong to either upper or lower 10% of engineering classes in the universities or other similar institutions, question problems and new ideas, have fewer close friends, think in more abstract and theoretical terms, are more stable, display more self-confidence, possess a high desire for freedom, and face ambiguous situations more easily [10, 11].

Creativity among managers is also very important to the success of any organization. The typical characteristics of creative managers are creative memory; self-motivation; self-competition; fearlessness of authority; fearlessness of failure; concentration power; respect for others' rights; sensitivity to problems; originality and self-confidence; a lack of jealousy; ability to toy with ideas; persistence; tolerance of ambiguity; ability to analyze and synthesize; active curiosity; openness to feelings; optimistic; nonauthoritarian, open, and direct when dealing with others; and a lack of resentment toward authority [1, 11].

Over the years researchers involved with creativity have also studied individuals from noncreative aspects and established the following main characteristics of noncreative persons [11, 19].

- Cynical;
- Fear of ridicule;

- Resistant to change;
- Conformity;
- Fear of failure;
- No interest in experimenting;
- Inclination towards systematic routine;
- Seeking of security;
- Jealous of competitiveness;
- Nonbelief in nonconventional ideas.

7.6 New Idea Generation, Presentation, Evaluation, and Elimination

Each of these items is discussed here, separately.

7.6.1 New Idea Generation

Although, there is no magic formula to generate new ideas, the general actions such as those listed next help to generate new thoughts [11, 17]:

- Choose a most useful time of the day to think of new ideas;
- Set a deadline to generate new ideas;
- Establish a quota for the ideas;
- Record ideas;
- Use checklists;
- Employ creative questioning;
- Use other aids as appropriate.

7.6.2 New Idea Presentation

This is probably the most important aspect of the creative process because even the most brilliant idea may die if its presentation is not effective. In general, the idea presentation guidelines include obtaining information on the audience, selling the new idea to your boss, obtaining information on conditions that facilitate idea selling, looking for the right timing when selling new ideas, and selling by implanting.

In the actual presentation of new ideas, some of the useful guidelines are as follows [1, 11].

- Avoid arguments.
- Avoid unsubstantiated claims.
- Be prepared for objections.
- Add some touch of showmanship.
- Remain as natural as possible.
- Show only muted enthusiasm.
- Present problem background and the proposed solution approach.
- Present the immediate steps necessary for the new idea implementation.
- Remain calm and in good emotional control.
- Welcome suggestions to improve the proposed idea.
- Give an appropriate degree of attention to the practical aspects of the new idea.
- In the event of having technically or professionally sophisticated individuals among the audience, try to introduce some counter-arguments to the new idea and discuss them as considered appropriate.
- Communicate the new idea in concise, clear, and simple language.
- Purposely leave out some important details of the idea for others to fill in. This action will make others feel that they have also contributed to the idea. If no one fills in the gaps, fill them in by yourself at an appropriate time.
- Neutralize objections to the idea through means such as follows:
  - Listen carefully to the objecting individual and encourage him or her to elaborate on the objection. Past experience indicates that the more an individual talks, the weaker his or her argument against the idea becomes.
  - Ask such questions of the objecting individual with which he or she has to agree.
- Arrange for a break after the presentation. This exercise allows time for the audience to sort out questions concerning the presentation. After the scheduled break, try to give answers to all questions arising from the audience.

At the end, sum up the presentation and benefits of the new idea and the important reasons for its implementation.

7.6.3 New Idea Evaluation

Usually, a new idea has to pass through an evaluation process prior to its implementation. New ideas may be grouped under the following three classifications [1, 11]:

- *Simple and straightforward ideas.* These ideas do not require much effort or expenditure, and they can be acted upon immediately.
- *Difficult ideas.* These ideas can only be accomplished in a year or more time and require an extensive effort.
- *Moderate ideas.* These ideas are likely to be accomplished within 6 months with a reasonable investment.

Different organizations may use varying factors to evaluate a new idea. A chemical company made use of factors such as growth, research and development, marketability, stability, engineering, production, and the present company position to evaluate new product lines [1, 11]. More specifically, the company evaluated each new product idea with respect to all of these factors and then rated it into any one of the five distinct categories: very good, good, average, poor, and very poor.

7.6.4 New Idea Elimination

Management plays an instrumental role in the creativity of the personnel it manages. More specifically, the creativity effort of individuals will only be effective if their manager is receptive to their new ideas. In fact, the negativity of a manager may be killing good ideas that may be worth millions of dollars someday. Past experience indicates that there are many negative ways a manager may act towards members of his or her department who come up with a new idea. Some examples of these ways are as follows:

- The idea is too radical.
- The idea is too simple.
- The idea is too complex.
- The idea will cost millions of dollars.
- The idea is too theoretical.
- It is a ridiculous idea.
- Our organization is doing quite well without it.

- The higher management will not buy it.
- Our organization is too big or too small to try this idea.
- The idea is outside our jurisdiction.
- We do not have sufficient time to try this idea.
- Our current budget would not be able to accommodate it.
- We have already tried similar ideas.
- Let's form a committee.

Responses such as these, coming from a manager, may very well be killing creativity within a group. More specifically, a good manager always avoids such negative responses to obtain healthy creativity results from his or her subordinates.

7.7 Creativity Methods

Over the years, researchers working in the field of creativity and inventiveness have developed many methods to generate new ideas. According to some professionals, there are over 30 creativity methods that can be used to find solutions to various types of problems. Some of these methods are presented next [11, 15].

7.7.1 Group Brainstorming Method

This is perhaps the most widely used method to generate new thoughts in the industrial sector. In modern context, Alex Osborn [8] was the first person to apply it. As per Osborn, the approach was practiced by Hindu religious teachers for over 400 years, and it was called Prai-Barshana: Prai means "outside yourself" and Barshana means "question." Consequently, the modern term *brainstorming* basically means "using the brain to storm a problem. "

A group of individuals participate in brainstorming sessions. During the sessions, one idea for finding a solution to a problem triggers another idea and the process continues. The individuals participating in the sessions belong to different backgrounds but have similar interests, and each brainstorming session is held for less than an hour and sometime for as short as 15 minutes. However, during these sessions the concentration on a given problem is very intense and the aim is to achieve at least 50 ideas in each session. Past experience indicates that the best results are obtained when 8–12 people participate in a session. Additional useful guidelines for conducting effective brainstorming sessions are as follows [11, 15, 23]:

- *Do not permit criticism.* This means no criticism whatsoever is allowed during sessions.

- *Welcome freewheeling.* The wilder the idea, the better it is. The main reason for this belief is that it is easier to discard ideas than to think them up.

- *Keep the ranks of participating individuals fairly equal.* Past experience indicates that it may take a substantial amount of warm-up time for low-ranking individuals to mix their thoughts freely with those produced by more senior individuals.

- *Record all ideas.* Record all ideas presented effectively during brainstorming sessions; otherwise they will be forgotten.

- *Select the timing of the session with care.* Ill-timed sessions may result in lower productivity of the participants subsequent to the sessions. For example, if a brainstorming session was held sometime during the morning, the participants may still be talking about the session later on during the day, thus leading to lower productivity.

- *Combine and improve ideas.* A single idea may not be a perfect solution to the problem, but combining it with others may turn out a better solution to the problem.

- *Think of some possible solutions to the problem under consideration ahead of time.* Past experience indicates the participants' idea output lags. During such periods, these solutions could be kicked in to stimulate the idea-generation process.

7.7.2 Checklist Method

This is one of the simplest techniques to generate solutions to a given problem. The checklist method simply calls for preparing a list of general questions on a given problem and then seeking answers to these questions. The method assumes that a solution to the problem exists. Examples of the checklist method questions are as follows [23–25]:

- Is it possible to modify the existing solution?
- Can the parts be rearranged?

- Is it possible to further improve the current benefits of the existing solution?

- Can the solution or feature be enlarged or reduced?

- Is there any other application of the current solution?

- Are there any other scientific bases that can equally be effective for the proposed solution?

- Is it possible to overcome the drawbacks of the existing solution?

- Can the existing solution be made more compact?

- What are the ways to further improve the quality, performance, and appearance of the existing solution?

- Are there any consequences if the solution is taken to its extreme?

- Is it possible to combine the solution with another solution to make it better?

7.7.3　CNB Method

This is a group-based method and assumes that the group members fully understand the problem objective and are willing to corporate. The method is composed of the following steps [16]:

- Provide each group member with a package containing items such as a notebook, problem description, preparation material, and relevant creative aids.

- Give each group member time (e.g., 1 week) to find a solution to the problem, and request that all ideas must be recorded each day of the allotted time period. At the end of the time period, require each person to select his or her best idea, summarize the remaining ideas, and write his or her thoughts in the notebook for further exploration.

- Gather all the notebooks, study all the proposed ideas with care, and prepare a report on the issue.

- Invite all members of the group to review all of the collected notebooks.

- Schedule a meeting of all participating persons to review the proposed solutions and choose the best solution.

7.7.4 Gordon Method

This method is named after W. J. Gordon [26] and it has been successfully applied to develop various types of new products (e.g., a gasoline pump, a can opener, and a razor). The basis for the method is the effort of a group, and usually the approach requires on average about 1 day of discussion. A total of six individuals participated in Gordon's team and the following two important characteristics are associated with the method:

- The entire team investigates the underlying concept of the problem. For example, in developing a new can opener, the entire team members would examine the concept of opening from all angles.
- The whole team investigates the subject from various different aspects. In the case of the can opener example, the group would first examine the meanings of the word "opening" and would then study various examples of opening. This is quite useful to discover unconventional approaches. After the full identification of these approaches, the team then develops the final approaches.

7.7.5 Forced-Relationships Method

This method was developed by C. S. Whiting, and its main objective is to produce new ideas by creating a forced relationship between two or more generally unrelated items or ideas [8]. This approach is demonstrated through an example. Assume that an office equipment manufacturer manufactures items such as a desk, chair, desk lamp, and bookcase. In this case, the starting point for the idea thinker could be the relationship between the desk and chair. From this point onward, the idea thinker would seek to start a line of free associations and ultimately he or she may click with an idea for a new product, for example, manufacturing a unit combining both desk and chair. At the end the most profitable or promising idea is chosen for further exploration or implementation.

7.7.6 Single-Person Method

In this method, only one person is involved and this individual documents the problem and then distributes it to participants prior to the meeting [27]. Furthermore, this individual specifies a certain number of solution ideas per participant as a requirement to participate in the meeting.

This exercise allows the individual to obtain a large number of tentative solution ideas. One of these ideas may end up as a sound solution to the problem.

7.7.7   Attribute-Listing Method

This technique was developed by R. P. Crawford for a single-person use and requires the listing of attributes of an idea or item [8]. After the completion of the listing, the individual considers each and every attribute at one time with the intention of making improvements on the item or idea to which these attributes belong. The method is described in detail in [8].

7.7.8   And-Also Method

This method was developed for use by two individuals [11, 16, 27]. Both of the individuals pick a subject for brainstorming and then agree with each other's idea, and then make some additions to other's idea. This process continues until an acceptable idea clicks.

7.7.9   Tear-Down Approach

This method is basically the same as the preceding approach but with one exception—instead of agreeing with each other, both persons tear down each other's idea [11, 16, 27]. More specifically, after picking a subject for brainstorming, person A takes the attitude that the current way is wrong and then contributes an idea suggesting another way. It is just another way but not necessarily better.

Person B is not permitted to agree with person A's idea. Consequently, person B has to come up with another way. In turn, person A is forbidden to agree with person B's way and has to suggest still another way. This process continues until some acceptable idea clicks.

7.8   Problems

1. Write an essay on the history of creativity.
2. Discuss the following terms:
   - Creativity;
   - Invention.

3. What are the seven classifications of inventions?

4. Discuss the ways to develop one's creative talent.

5. What are the creative problem-solving steps?

6. Discuss the following classifications of barriers to creative thinking:

- Cultural barriers;

- Emotional barriers.

7. What are the typical characteristics of a creative engineer?

8. What are the typical characteristics of a noncreative person?

9. List at least 10 useful guidelines for the presentation of new ideas.

10. Describe the following creativity methods:

- The group brainstorming method;

- The checklist method;

- The CNB method.

References

[1] Randsepp, E., *What the Executive Should Know About Creating and Selling Ideas,* Larchmont, NY: American Research Council, 1966.

[2] Rothenberg, A., and B. Greenberg, *The Index of Scientific Writings on Creativity,* Hamden, CT: Archon Books, The Shoe String Press, 1976.

[3] Albert, R. S., and M. A. Runco, "A History of Research on Creativity," in *Handbook of Creativity,* R. J. Sternberg (ed.), New York: Cambridge University Press, 1999, pp. 16–31.

[4] Mayer, R. E., "Fifty Years of Creativity Research," in *Handbook of Creativity,* R. J. Sternberg (ed.), New York: Cambridge University Press, 1999, pp. 449–460.

[5] Dieter, G. E., *Engineering Design,* New York: McGraw-Hill, 1983.

[6] Shannon, R. E., *Engineering Management,* New York: John Wiley and Sons, 1980.

[7] Kivenson, G., *The Art and Science of Inventing,* New York: Van Nostrand Reinhold, 1977.

[8] Osborn, A. F., *Applied Imagination,* New York: Charles Scribner's Sons, 1963.

[9] Tangerman, E. J., "Creativity: The Facts Behind the Fad," in *Management Guide for Engineers and Technical Administrators,* N. P. Chironis (ed.), New York: McGraw-Hill, 1969, pp. 262–265.

[10] Karger, D. W., and R. G. Murdick, *Managing: Engineering and Research,* New York: Industrial Press, 1969.

[11] Dhillon, B. S., *Engineering Management*, Lancaster, PA: Technomic Publishing Company, 1987.

[12] Van Frange, E., *Professional Creativity*, Englewood Cliffs, NJ: Prentice Hall, 1959.

[13] Bronikowski, R. J., "Creativity Steps," *Chemical Engineer*, July 31, 1978, pp. 103–108.

[14] Alger, J. R. M., and C. V. Hays, *Creative Synthesis in Design*, Englewood Cliffs, NJ: Prentice Hall, 1964.

[15] Dhillon, B. S., *Engineering Design: A Modern Approach*, Chicago, IL: Irwin, 1996.

[16] Haefele, J. W., *Creativity and Innovation*, New York: Reinhold Publishing Corporation, 1962.

[17] Mason, J. G., *How to Be a More Creative Executive*, New York: McGraw-Hill, 1960.

[18] Adams, J. L., *Conceptual Blockbusting: A Guide to Better Ideas*, 3rd ed., Reading, MA: Addison-Wesley, 1986.

[19] Harrisberger, L., *Engineermanship: A Philosophy of Design*, Belmont, CA: Wadsworth Publishing Company, 1966.

[20] McPherson, J. H., "Are You Creative?" *Prod. Eng.*, November 1958, pp. 28–29.

[21] McPherson, J. H., "The Relationship of the Individual to the Creative Process in the Management Environment," Paper No. 64MD12, American Society of Mechanical Engineers, New York, 1964.

[22] Dhillon, B. S., *Quality Control, Reliability, and Engineering Design*, New York: Marcel Dekker, 1985.

[23] Beakley, G. C., and E. G. Chilton, *Introduction to Engineering Design and Graphics*, New York: MacMillan, 1973.

[24] Walton, J., *Engineering Design: From Art to Practice*, New York: West Publishing Company, 1991.

[25] Stoecker, W. F., *Design of Thermal Systems*, New York: McGraw-Hill, 1980.

[26] Gordon, W. J., *Synectics*, New York: Harper and Brothers, 1961.

[27] Studt, A. C., "How to Set Up Brainstorming Sessions," in *Management Guide for Engineers and Technical Administrators*, N. P. Chironis (ed.), New York: McGraw-Hill, 1969, pp. 276–277.

8

Concurrent Engineering

8.1   Introduction

Concurrent engineering may be described as the simultaneous, interactive, and interdisciplinary involvement of people belonging to diverse backgrounds including design, manufacturing, and field support working together to reduce the product development cycle while ensuring factors such as reliability, performance, quality, and support responsiveness [1].

Past experience indicates that there are many benefits of concurrent engineering and integrated product development, including 65% to 90% fewer engineering changes, 30% to 70% less development time, 200% to 600% higher quality, 20% to 110% higher white-collar productivity, and 20% to 90% less time to market [2, 3].

Although concurrent engineering has been around in one form or another for a very long time, its modern form may be attributed to the 1980s when the Ford Motor Company practiced the team, or concurrent engineering, approach in the design and development of its Taurus model [4–8]. In 1982, the Defense Advanced Research Projects Agency (DARPA) initiated a project to develop ways and means to improve concurrency in the design process and in 1987 the final results of this study were released [9].

In 1986, the term *concurrent engineering* was coined by the Institute for Defense Analyses in its report R-338 [10]. The following year, DARPA formed a working group composed of experts from government, industry, and academia to evaluate the implications of simultaneous engineering for defense sourcing [4]. The group wholeheartedly supported the concept of *simultaneous engineering* but rechristened it to *concurrent engineering*.

By the end of 1991, the U.S. government had allocated around $60 million under the auspices of the DARPA initiative for developing current engineering tools and other areas [4, 11]. The introduction of concurrent engineering to industrial sectors such as defense, aerospace, and automobile acted as a trigger at the beginning of the industrial supply chain, and suppliers and subcontractors played an instrumental role in rapidly spreading its use.

This chapter presents various aspects of concurrent engineering.

8.2   Concurrent Engineering Objectives and Basic Principles

The objective of improving product quality is very important, as today's customer is becoming very quality conscious. There could be many objectives for practicing concurrent engineering in an organization and some of the important ones are listed in Table 8.1 [8, 12]. Each of these objectives is discussed next.

Consequently, in the future the better quality products are likely to take the lion's share of the market. As the application of concurrent engineering helps to improve product quality, the primary objective of some companies for practicing concurrent engineering is to improve the quality of their products. For example, Hewlett-Packard practiced concurrent engineering to enhance the quality of its products by 100% in a given time [13].

As the development cost is an important element in the selling price of a product, its reduction can help to improve the competitive advantage of the manufacturer.

Similarly, the manufacturing cost is a significant component of the total product cost, and the practice of concurrent engineering can help to reduce the manufacturing cost by producing manufacturing-friendly product designs.

Reducing marketing time basically means responding faster to the customer requirements, and the concurrent engineering approach is a useful tool for achieving this.

Improving the competitiveness of manufactured products is vital, as the competition in today's world increases globally. Many manufactured products have to compete with internally and externally manufactured products. Consequently, the practice of concurrent engineering is not only important to increase the market share but also to maintain it, because the

Table 8.1
Main Objectives of Concurrent Engineering

Number	Objective
1	Enhance product quality
2	Reduce product development cost
3	Reduce manufacturing cost
4	Reduce marketing time
5	Improve manufactured products' competitiveness
6	Reduce cost of testing
7	Increase profit margins
8	Reduce service cost

manufacturers of the competitive products could very well be taking advantage of concurrent engineering.

Reducing the cost of testing is also important. As many manufactured products continue to increase in complexity and sophistication, the cost of testing is becoming a larger component of the overall product cost equation. For example, the cost of testing may increase due to the incorporation of built-in test systems in such products. Concurrent engineering is a useful tool to reduce such cost.

Increasing profit margins is a major benefit—usually organizations in business give a very high importance to profit margins and employ concurrent engineering to achieve their goals concerning profit.

Reducing service cost is beneficial, as many large-size products require some form of servicing after their installation at the customers' facility. The concept of concurrent engineering is a useful tool to direct attention during product design to lower serviceability-related costs.

The foundation of concurrent engineering is laid on eight basic principles shown in Figure 8.1: feeling of ownership, mutual understanding, early problem discovery, team effort affinity, work structuring, early decision making, constancy of purpose, and knowledge leveraging [14].

The feeling of ownership means the team members will work effectively to produce a good product if they are provided with the authority to shape its design as considered appropriate. Mutual understanding means the team members will work better when they are aware of what other members are doing.

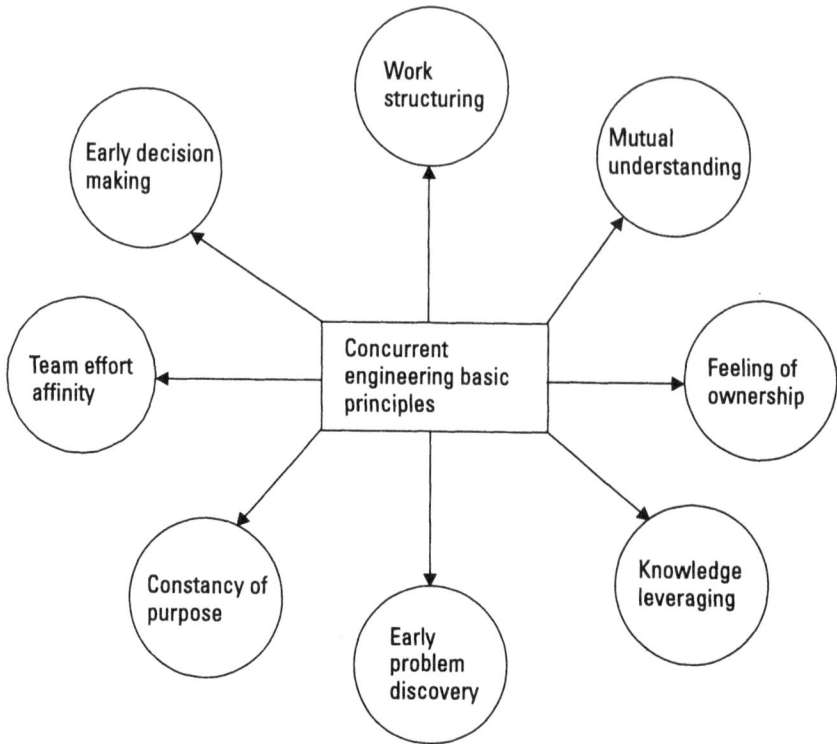

Figure 8.1 Fundamental principles of concurrent engineering.

Early problem discovery means the problems found early in the design phase are easier to solve than the ones found at a later stage. Team-effort affinity means teams will have a better affinity when there is trust among teams. Also, past experience indicates that an optimal product design produced by an individual team seldom remains optimal in a combined domain (i.e., in the environment of multiple team efforts).

Work structuring means that from the practical aspect, unlike parallel computers, human minds cannot work simultaneously on multiple tasks. However, the human mind is quite good at systematically structuring the work or structuring the work environment so that an individual task can be executed independently either by a machine or human. Early decision making means the window of opportunity to affect a given design is much wider during early stages of design than the later ones. Furthermore, past experience indicates that often teams have a natural tendency to make good and efficient decisions.

Constancy of purpose means everyone involved contributes his or her best when working towards a common and consistent goal. Knowledge leveraging means that as the domain of product design is often quite large, interlinking decision support tools with spurts of human knowledge base is the most viable approach to find solutions to complex problems.

8.3 Applying the Concurrent Engineering Concept

The application of the concurrent engineering concept may not be beneficial to all companies. These factors dictate the degree and the need for concurrent engineering in an organization:

- Size of organization;
- Type of product;
- Process used to produce the product;
- Physical locations of the company branches.

In order to develop an overall vision of a concurrent engineering environment in a company, careful attention must be paid to the following four areas [8, 9]:

1. *Product-development environment.* This is concerned with assessing the product-development environment of an organization. Some of the available tools for this purpose are the Department of Defense CALS/CE Task Group for Electronic Systems Self Assessment, the Mentor Graphics Corporation Process Maturity Assessment Questionnaire, and the Carnegie Mellon University Software Engineering Institute Assessment Questionnaire [9]. Another tool that combines elements from tools such as these listed here is described in [9] and is made up of four major parts: the company-assessment questionnaire, the methods matrix, the dimensions map, and the priority road map. All of these tools are useful for examining the existing product-development environment within the company, suggesting a vision for the right kind of concurrent engineering environment for the product to flourish, illustrating the company needs for bringing the dimensions of concurrent engineering into balance, and planning a road map for transforming the existing product-development environment into a concurrent engineering one.

2. *Current standing.* This is concerned with determining the current standing of the organization with respect to product development. This can be accomplished through a company-assessment questionnaire concentrating on four areas: organization, communication infrastructure, requirements, and product development. The organization-related questions address areas such as team integration, training, automation support, and empowerment. The communication infrastructure–related questions address the effective functioning of the communication infrastructure with respect to product development, product data, and feedback. The requirements-related questions address the issue of effectiveness in meeting internal and external requirements covering such areas as planning methodology, planning perspective, requirements definition, and planning perspective, validation, and standards. The product development–related questions address areas such as design process, optimization, and component engineering.

3. *Envisioned objective achievement approaches.* This is basically concerned with determining the methods that can be effectively employed in product-development environments to attain the maximum benefits of concurrent engineering. For this purpose, past experience indicates that the methods-matrix approach can be a very effective tool. This approach is described in [9].

4. *Plan to achieve set objectives.* This is concerned with developing plans to achieve set goals. These goals should be developed by using the results of the company-assessment questionnaire and the methods-matrix approach. Actions such as the following are appropriate:

- Plot assessment results;
- Perform analysis of the results;
- Communicate and make refinements to the concurrent engineering vision;
- Validate with a pilot;
- Make improvements continuously.

During the consideration for the application of the concurrent engineering approach, the four important factors that require careful consideration are concurrence, constraints, coordination, and consensus [15].

8.3.1 Concurrent Engineering Concept Introduction—Important Related Factors

During the introduction of concurrent engineering, careful consideration to many factors is absolutely essential. Some of the important general factors to which answers should be sought are listed here [15].

- Start date and the location of the concurrent engineering approach;
- Management's expectations from the project;
- Procedures to be followed in weighing conflicting objectives;
- Ways and means to be followed in evaluating results of the concurrent-design project;
- Degree of reliance on external concurrent engineering–related expertise;
- Method to be used in managing the concurrent engineering team;
- Reporting of the team within the organization and its members' physical location;
- Training needs of the team members to function as a group;
- Procedure to be employed to ensure team productivity.

8.4 Concurrent Engineering Team and Associated Areas

The cornerstone of the concurrent engineering approach is the use of teams. These teams are very collaborative in nature, with each member using his or her expertise and resources to enhance the team goals. Some of the objectives of the collaborative team approach within concurrent engineering include cooperation intensification, establishing an atmosphere that fosters effective communication, making the functional team members responsible for checking for possible product and business risks and avoiding downstream pitfalls, and allowing for early unfinished releases of the design to be examined by the team members.

The team-cooperation philosophy is composed of seven elements: collaboration, communication, compromise, coordination, commitment, consensus, and continuous improvement [14].

8.4.1 Concurrent Engineering Team Formation

For the success of the application of the concurrent engineering concept, careful consideration is needed during team formation. For example, it is not only essential to include team members from all concerned areas and disciplines, but also to ensure that they all have appropriate qualifications, experience, and interpersonal skills.

A typical concurrent engineering team includes members such as a concurrent engineering mentor, team leader (engineering), design engineer, information-technology specialist, engineering manager (lead), vendor or customer representative, marketing manager, quality-control engineer, service engineer, manufacturing engineer, human factor or environmental specialist, safety engineer, reliability engineer, and software engineer [16].

The team members should represent all involved disciplines, possess a broad view, believe in the company's success, have a strong team spirit, possess adequate capability for compromising and accepting consensus decisions, and possess some knowledge about other disciplines.

A study conducted in 1990 revealed that the typical component-design team in the U.S. electronics industry was composed of an average of eight people as opposed to 18 in Japan [17].

8.4.2   Concurrent Engineering Team Plan

This is an important document of concurrent engineering and includes the major components of the activity. The important categories that should be used at the initial stage of the development cycle include team charter and membership, preliminary development schedule, applicable standards, key competing products, projected product volume and cost market requirements, applicable regulatory requirements, preliminary business projections, customer need projections, projection of competitors' market positioning, manufacturing risk or impact, key competitive advantages, development budget needs, product distribution requirements, key technologies to be used, competitive market position, and a customer information collection process [16].

8.4.3   Concurrent Engineering Team Charter

It is essential to have a clearly written team charter. The primary objective of having the charter is to avoid confusion or misunderstanding. The charter should include items such as goal and purpose of the team, description of the product under consideration, decision making, budget authority, implementation requirements, each team member's role, the team membership, team authority and its boundaries, the senior manager accountable for the team, the competitive situation, and support from the company president and/or other senior management people.

8.4.4 Concurrent Engineering Team Leadership and Mentor's Role

Usually, the concurrent engineering team leader belongs to the engineering department. He or she performs tasks such as calling the team meeting, chairing the meeting, managing the budget, holding the review meetings, managing the project schedule, developing the plans, organizing the team, resolving conflicts, coordinating the team activities, and recruiting new members. Also, the concurrent engineering team leader is management's main contact with the product team.

The leader should possess qualities and traits such as those listed here [17, 18]:

- Flexibility and honesty;
- Ability to reason and be tactful with lead team members;
- Good interpersonal skills and the ability to motivate team members;
- Good memory and the wisdom to avoid repeating the same mistakes;
- Ability to assign tasks to team members in a pleasant manner;
- Aptitude to develop the agenda for each meeting;
- Ability to recognize differing views;
- Ability to resolve important issues through consensus;
- Ability to ensure the distribution of essential information in an effective manner to all concerned;
- Knowledge to garner appropriate resources such as hardware, software, and human;
- Strong participation in outlining the problems to be solved.

The role of the concurrent engineering mentor is situational and is particularly important for organizations new to concurrent engineering application. This individual possesses previous experience in the implementation of concurrent engineering and is in a position to help the team convert its old style of thinking into the new approaches of concurrent engineering. The main advantage of having a concurrent engineering mentor is to get the team fully into the concurrent engineering mindset faster. Some of the functions performed by this person are as follows [16]:

- Train team members in concurrent engineering approaches.
- Elevate systemic problems inhibiting the concurrent engineering process.

- Document lessons learned and pass them on to all concerned.

- Work with team members and managers on concurrent engineering matters.

8.4.5   Team Management and Limitations

Over the years, much has been written on managing teams. Some important guidelines highlighted by the professionals in the field concerning team management are as follows [16, 18]:

- Develop team goals.

- Maximize team collaboration and organizational support.

- Establish trust as much as possible among team members.

- Ensure that the team performs its assigned function under its own leadership.

- Ensure that the team meetings are held regularly.

- Conduct brainstorming sessions as appropriate among team members to overcome difficulties.

- Measure and review team contributions regularly.

- Prepare clear team meeting agendas and distribute them to all concerned individuals 3 to 5 days before the meeting.

- Document the minutes of the meetings and distribute them to all concerned individuals as soon as possible after each meeting.

There are various limitations associated with the teams, including general team limitations, management limitations, and team self-imposed limitations. Two examples of the general team limitations are slower team processes (i.e., relative to individual processes) and difficulty recognizing the contributions of individual members to the result. The management limitations are concerned with team limitations imposed by the management that affect the team operation. Team self-imposed limitations are concerned with factors such as the lack of recognition of expertise, large teams, confusion between roles and expertise, and inappropriate background to make the required contributions.

8.4.6   Good Concurrent Engineering Team Characteristics

There are many characteristics associated with a concurrent engineering team performing its tasks effectively and some of these are as follows [15]:

- Clearly defined and agreed-upon goals;
- Well-defined process and procedures;
- Dynamic leadership;
- Clearly defined functions of team members;
- Excellent cooperation among team members;
- Positive relationships among team members;
- Existence of healthy and low levels of conflicts;
- Open and frank communication among team members.

8.5   Concurrent Engineering Process–Related Methodologies and Techniques

Over the years, many methodologies and techniques have been developed that can also be used in the concurrent engineering process. In fact, a significant part of the concurrent engineering process is concerned with proper and efficient use of important methodologies and techniques.

Many methodologies and techniques can be used in the concurrent engineering process, including quality loss function, *quality-function deployment* (QFD), *design for manufacturing* (DFM), Pugh process, *design-stress analysis* (DSA), Taguchi's robust design approach, signal-to-noise ratio, experiment by orthogonal arrays, rapid prototyping, *customer-focused design* (CFD), benchmarking and competitive analysis, and Ishikawa's seven tools (i.e., Pareto diagram, cause-and-effect diagram, histogram, binomial probability paper, control charts, scatter diagram, and check sheets) [16, 19].

Some of these methodologies and techniques are described next [16, 19].

QFD.   This is also referred to as the *house of quality*. The approach was devised by Professor Yoji Akao and applied first by two Japanese companies: Toyota Motor Corporation and Mitsubishi Corporation. In the late 1980s, the QFD method was first introduced in North America, and it may simply be described as a pair of spreadsheets relating subjective customer desires, called *customer attributes*, to quantitative engineering characteristics. QFD provides a justification for establishing key product features and the projected price of the product under consideration. This method is used during the conceptual level of the design process and is extremely useful in eliminating low-priority product features.

Pugh process. Just like QFD, this is a matrix methodology and is useful for comparing various alternatives, in particular identifying the better attributes out of many alternative attributes [20]. It permits the determination of each alternative on the basis of a preestablished criterion. Furthermore, past experience indicates that the Pugh approach is a useful tool for the design team to gain a greater look at the strengths and weaknesses of all proposed possible solutions, thus it is a powerful approach to choose alternative concepts and designs.

Taguchi's robust design approach. This is a powerful method with its own terminology and was developed by Genichi Taguchi at the Electrical Communications Laboratory of the Nippon Telegraph and Telephone Company in Japan during the period 1949–1961. This method views design as a system having inputs, outputs, noise factors, and control factors. Furthermore, it aims to minimize the deviation from the desired target value (e.g., of a power supply output voltage) while keeping the manufacturing cost at its minimum level. The deviations addressed by the approach are typically due to environment (e.g., heat transfer), material variations (e.g., capacitance variation), and manufacturing process (e.g., assembly errors or variations). All in all, the main objective is to minimize such deviations' impact on the designed function.

DFM. This method evolved in the United States over a period of time, and it originally began as a producibility function with the goal of ensuring that a given design could be produced in volume in manufacturing. DFM may simply be called a method of design with the objective of understanding the future manufacturing processes of the product during the design phases. Consequently, the team can maximize manufacturing quality and minimize manufacturing costs. Although, DFM approach is multifaceted, it can essentially be broken into three parts (i.e., A, B, and C). Part A applies to the concept design, part B to the detailed design, and part C is concerned with final adjustments just before the production.

DSA. This approach is used as an early indicator of design-related problems, and it employs methods of accelerating early failures and combines these with parametric stresses. The primary objective of DSA is to help in understanding design and selected components' weaknesses, thus eradicating potential failure modes prior to their existence in the final product. DSA may simply be called a design-reliability and quality-improvement activity and is best employed in complex designs and in situations where reliability between design variations needs to be understood effectively.

Rapid prototyping.   Over the years, this methodology has emerged as a bona fide approach for clearly understanding the specifics of the design. The early prototypes are produced for validating appearance, demonstrating concepts, and providing insights into designs. In the past, rapid prototyping has proven to be good for early customer feedback and feedback from the manufacturing people.

Cause-and-effect diagram.   This method belongs to one of Kaoru Ishikawa's seven tools, and he developed it in the early 1940s. This approach is also referred to as the *Ishikawa diagram* or the *Fishbone diagram*. The reason for calling it the Fishbone diagram is that it resembles the bones of a fish. The right side of the diagram (i.e., the fish head) represents the effect and the left side represents all possible causes connected to the central "fish" spine. More specifically, the diagram starts with an effect as a spine and progresses in a backward direction with each major category of causes added as ribs. In turn, specific causes are added as branches on the ribs. The advantages of this approach include its use as a tool to determine root cause, generate ideas, guide further inquiry, and present a systematic arrangement of theories.

CFD.   This approach is useful for providing a significant amount of customer input into the design process, thus requiring the construction of samples and models for the examination of customers to gauge their inputs. In turn, the design team reviews customer input and suggests appropriate measures. The process continues until customer satisfaction is reached.

8.6 Useful Concurrent Engineering Guidelines and Guidelines for Handling Conflicts in the Concurrent Engineering Environment

Over the past decade or so, professionals working in the concurrent engineering area have developed various useful guidelines. Some of these guidelines are as follows [8, 12, 21]:

- Form a multidisciplinary design team.
- Prepare effectively for concurrent engineering implementation.
- Make allowances for a concurrent engineering learning curve.
- Establish effective communication with product users or customers.
- Integrate technical reviews in an effective manner.
- Involve representatives of suppliers and subcontractors during the early phases of the project.
- Benefit from lessons learned or experience gained.
- Integrate with care computer-aided engineering tools with the product model.
- Perform simulation analyses of product or process performance(s).
- Design associated processes concurrently with the item or product under consideration.
- Improve the design process continuously.
- Implement concurrent engineering in manageable bites.

Just as with any other engineering project, from time to time conflicts arise in the concurrent engineering environment (e.g., between the requirements and desires of individual concurrent engineering team members). Wise concurrent engineering leadership plays a key role in turning such conflicts into healthy dialogues. Some important guidelines for handling conflicts in the concurrent engineering environment are listed in Table 8.2 [12, 21].

Table 8.2
Important Guidelines for Handling Conflicts in Concurrent Engineering Environment

Number	Guideline
1	Find out reasons for conflicts
2	Always shoot for consensus decisions
3	Place all requirements of the customers at the forefront
4	Handle with care the input of each team member
5	Compromise as much as possible

8.7 Concurrent Engineering Advantages, Savings Due to Concurrent Engineering, and Its Risks

The practice of concurrent engineering helps to generate many advantages, including shortening the time and lessening the cost for products to get to market, inherent quality in product design, many processes occurring concurrently rather than sequentially, reduction in the learning curve because of cross training, better use of scarce technical resources, reduction in lost time caused by communication breakdowns, decreased occurrence of obsolescence, preemption of errors and early problem detection, creation of flexibility to accommodate changes, and provision of the best overall input [14].

Over the years many companies in the United States have employed the concurrent engineering approach and reported savings in the terms of cost, quality, and time as presented in Tables 8.3–8.5, respectively [12, 22, 23].

Table 8.3

Savings in Cost Due to Concurrent Engineering in Various U.S. Companies

Number	Company Name	Cost-Related Savings (Direct/Indirect)
1	Hewlett-Packard: Instrument Division	42% reduction in manufacturing costs
2	Northrop	Approximately 30% savings on bid on a major-ticket item/product
3	AT&T	At least 40% reduction in the cost of repair for new circuit pack production
4	IBM	45% reduction in product direct assembly labor hours
5	McDonnell Douglas	Approximately 60% in savings on bid for reactor and missile projects
6	Deere and Company	Approximately 30% reduction in development cost for construction equipment
7	Boeing Ballistic Systems Division	Reduction in labor rates by $28 per hour; reduction in cost by 30% to 40%
8	NCR	44% reduction in manufacturing costs with respect to NCR's 2760 electronic cash register
9	Cisco Systems	Revenue increased from $27 million in 1989 to $70 million in 1990

Table 8.4
Savings in Terms of Quality Due to Concurrent Engineering in Various U.S. Companies

Number	Company Name	Quality-Related Savings (Direct/Indirect)
1	AT&T	Reduction in defects by 30% to 87%
2	Deere and Company	Reduction in number of inspectors by 66%
3	Northrop	Reduction in defects by 35%; reduction in number of engineering changes by 45%
4	IBM	Fewer engineering changes; guaranteed producibility and testability
5	McDonnell Douglas	Reduction in scrap, nonconformance, and weld defects per unit by 58%, 38%, and 70%, respectively
6	Hewlett-Packard: Instrument Division	Reduction in product field failure rate; scrap and rework by 60% and 75%, respectively
7	Boeing Ballistic Systems Division	Reduction in floor inspection ratio, material shortages, and defect-free operation by 66%, 12% to 0%, and 99%, respectively

Table 8.5
Savings in Time Due to Concurrent Engineering in Various U.S. Companies

Number	Company Name	Time-Related Savings (Direct/Indirect)
1	Deere and Company	Approximately 60% reduction in development time
2	Hewlett-Packard: Instrument Division	35% reduction in development cycle time
3	AT&T	Reduction in total process time to 46% of baseline for 5ESS
4	IBM	Approximately 40% reduction in electronic design cycle
5	McDonnell Douglas	Reduction in time from 45 weeks to 8 hours in one phase of high-speed vehicle preliminary design
6	Northrop	50% reduction in part and assembly schedule on two major subassemblies
7	Boeing Ballistic Systems Division	30% reduction in part and materials lead time
8	NCR	80% reduction in parts and assembly time with respect to NCR's 2760 electronic cash register
9	Texas Instruments	85% reduction in assembly time of a complex infrared sight

Although there are many advantages of concurrent engineering, it also has a number of associated risks. Some of those are listed next [14, 15].

- A good engineer, designer, or any other associated individual may not turn out to be a good team player.
- The degradation of turnaround time may occur, if the company in question is not properly equipped to process a large volume of information simultaneously.
- The iterative costs can increase under certain circumstances.
- The team-formation cost could be quite significant if design, production, and key staff functions are located at different places.
- Past experience indicates that it is rather more difficult to manage teams than individuals.
- During the implementation phase of the concurrent engineering approach, resistance to change can occur.

8.8 Problems

1. List at least eight objectives of concurrent engineering.
2. What are the fundamental principles of concurrent engineering?
3. Discuss the four areas to which careful attention must be paid in developing an overall vision of a concurrent engineering environment.
4. Discuss the important concurrent engineering concept introduction–related factors.
5. Discuss the following:
    - Concurrent engineering team plan;
    - Concurrent engineering team charter.
6. Describe the qualities and traits of a good concurrent engineering team leader.
7. List at least eight characteristics of a good concurrent engineering team.
8. Describe these concurrent engineering process–related methodologies:
    - Quality function deployment;
    - Pugh process;
    - Design for manufacturing.

9. List guidelines for handling conflicts in the concurrent engineering environment.

10. Discuss the benefits and risks associated with concurrent engineering.

References

[1] Sanchez, J. M., J. W. Priest, and L. J. Burnell, "Design Decision Analysis and Expert Systems in Concurrent Engineering," in *Handbook of Design, Manufacturing, and Automation*, R. C. Dorf and A. Kusiak (eds.), New York: John Wiley and Sons, 1994, pp. 51–63.

[2] Society of Concurrent Engineering (SOCE), Seattle, WA, 2001.

[3] *Business Week* (Benefits as reported by the National Institute of Standards and Technology), Thomas Group, Inc., and Institute for Defense Analyses, New York, April 30, 1990.

[4] Brookes, N., and C. Backhouse, "Concurrent Engineering: Where It Has Come From and Where It Is Now," in *Concurrent Engineering*, C. Backhouse and N. Brookes (eds.), Aldershot, England: Gower Publishing Limited, 1996, pp. 1–22.

[5] Smith, R. P., "The Historical Roots of Concurrent Engineering Fundamentals," *IEEE Trans. on Engineering Management*, Vol. 44, 1997, pp. 67–78.

[6] Ziemke, M. C., and M. S. Spann, "Concurrent Engineering's Roots in the World War II Era," in *Concurrent Engineering: Contemporary Issues and Modern Design Tools*, H. R. Parsaei and W. G. Sullivan (eds.), London: Chapman and Hall, 1993, pp. 24–41.

[7] Belson, D., "Concurrent Engineering," in *Handbook of Design, Manufacturing, and Automation*, R. C. Dorf and A. Kusiak (eds.), New York: John Wiley and Sons, 1994, pp. 25–34.

[8] Dhillon, B. S., *Advanced Design Concepts for Engineers*, Lancaster, PA: Technomic Publishing Company, 1998.

[9] Carter, D. E., and B. S. Baker, *Concurrent Engineering: The Product Development Environment for the 1990s*, Reading, MA: Addison-Wesley, 1992.

[10] Report No. R-338, Institute for Defense Analyses, Alexandria, VA, 1988.

[11] Reddy, R., R. T. Wood, and K. J. Cleetus, "The DARPA Initiative: Encouraging New Industrial Practices," *IEEE Spectrum*, July 1991, pp. 26–30.

[12] Turimo, J., *Managing Concurrent Engineering*, New York: Van Nostrand Reinhold, 1992.

[13] Shina, S. G., *Concurrent Engineering and Design for Manufacture of Electronics Products*, New York: Van Nostrand Reinhold, 1991.

[14] Prasad, B., *Concurrent Engineering Fundamentals*, Upper Saddle River, NJ: Prentice Hall, 1996.

[15] Bralla, J. G., *Design for Excellence,* New York: McGraw-Hill, 1996.

[16] Salomone, T. A., *Concurrent Engineering,* New York: Marcel Dekker, 1995.

[17] Askin, R. G., and M. Sodhi, "Organization of Teams in Concurrent Engineering," in *Handbook of Design, Manufacturing, and Automation,* R. C. Dorf and A. Kusiak (eds.), New York: John Wiley and Sons, 1994, pp. 85–105.

[18] Dhillon, B. S., *Engineering Management,* Lancaster, PA: Technomic Publishing Company, 1987.

[19] Hall, D., "Concurrent Engineering: Defining Terms and Techniques," *IEEE Spectrum,* July 1991, pp. 24–25.

[20] Pugh, S., *Total Design Integrated Methods for Successful Product Engineering,* Reading, MA: Addison-Wesley, 1991.

[21] Engineering Department Management and Administration Report (EDMAR), Institute of Management and Administration Inc., New York, May 1997.

[22] Dhillon, B. S., *Engineering Design: A Modern Approach,* Chicago, IL: Irwin, 1996.

[23] Zangwill, W. I., "Concurrent Engineering: Concepts and Implementation," *IEEE Engineering Management Review,* Vol. 20, No. 4, 1992, pp. 40–52.

9

Value Engineering

9.1 Introduction

The term *value* may mean different things to different people because it is used in a variety of ways. Around 350 B.C., Aristotle, the teacher of Alexander the Great, identified seven classes of value: economic, moral, social, religious, aesthetic, judicial, and political [1]. Interestingly, these classes are still recognized today. Nonetheless, value engineering may simply be described as an organized effort directed at analyzing the function of an item or a product with the objective of achieving the specified function at the lowest possible overall cost [2].

Past experience indicates that the application of the value engineering concept has helped to reduce manufacturing and procurement costs by approximately 15% to 25%. In particular, the U.S. Department of Defense reports that the use of value engineering has consistently produced a return on investment of 10:1 across the organization [3].

The history of value engineering may be traced back to 1947, when General Electric (GE) management selected Lawrence D. Miles, a GE electrical engineer, to devise appropriate approaches or methods that would be helpful in generating tangible savings through material or part substitutions, changes in manufacturing techniques, or design [4, 5]. Miles and Harry Erlicker, GE vice-president for purchasing, used the term *value analysis* [2]. In 1954, with the help of Miles and his GE team, the U.S. Navy Bureau of Ships set up a formal Navy value program and Rear Admiral Wilson D.

Leggett renamed the term *value analysis* to *value engineering* [6]. By 1956, all 11 naval shipyards were active in value engineering studies [7].

In 1959, the Electronics Industries Association organized the first national conference on value engineering at the University of Pennsylvania in Philadelphia, and the Society of American Value Engineers (SAVE) was formed [4]. In 1962, SAVE initiated the publication of a magazine entitled *The SAVE Journal of Value Engineering*. During the period 1963–1966, the U.S. Department of Defense reported the savings of over $1.1 billion due to the application of the value engineering concept [2].

In 1983, the Japanese established the Miles Award for companies that were most successful in reducing costs while keeping high quality of their products.

Over the years many publications on value engineering have appeared, and a list of most of the books published on the topic is given at the end of the chapter [1, 2, 4, 6–18]. This chapter presents various important aspects of value engineering.

9.2 Value Engineering Terms and Definitions

This section presents selective terms and definitions directly or indirectly concerned with value engineering [1, 2, 10, 13].

- *Value engineering.* This is an organized effort directed at analyzing the function of an item or a product with the objective of achieving the specified function at the lowest possible cost.

- *Value analysis.* This is a systematic approach for the efficient identification of unnecessary costs.

- *Value engineer.* This is an individual engaged in applying or promoting value engineering–associated ideas.

- *Value objective.* This is the same or enhanced performance at lesser cost.

- *Value engineering proposal.* This is a document that effectively provides details of a recommended change that will result in an overall cost improvement (i.e., a change in approach, equipment, or design).

- *Value improvement.* This is the improved function-to-cost ratio due to the application of the value-engineering approach to all product or service phases and their development.

- *Product.* This is a set of deliverables that are appropriate to satisfy the needs of customers.

- *Function.* This means different things to different people (e.g., role, duty, office, and faculty).

- *Unnecessary cost.* This is the total cost of items that do not really make any contribution to the essential functions, quality, reliability, or maintainability.

- *Savings.* This is the difference between original cost and cost that occurs after a change.

- *Cost prevention.* This is the elimination of unwarranted costs in the development stages of designs or operations.

- *Value engineering task force.* This is a group of value-oriented individuals with a clearly stated short-term task for developing, recommending, and implementing solutions for a cost or value problem.

9.3 Value Engineering Objectives and Reasons for the Unnecessary Cost and for Not Having a Value Engineering Program

The overall objective of value engineering may simply be described as providing a means of total cost control anywhere within an item's life cycle. It stresses cost reduction or elimination while maintaining the required quality and reliability of the item under consideration. The main goals of a modern value analysis may be stated as follows [6]:

- To eliminate or reduce the product, item, or process cost;

- To improve customer acceptance of the item, product, or process under consideration;

- To establish value analysis as an ongoing activity that will subsequently be applied to all company problems concerning cost or function.

It may be said that unnecessary costs are the ones that fail to contribute effectively to the product or item to which they accrue. Reference [1] has categorized reasons for unnecessary costs into three basic areas as shown in Figure 9.1. Mental conditioning includes five basic conditions (i.e., lack of information, honest wrong beliefs, habits and attitudes, temporary circumstances, and lack of ideas). Similarly, the two basic conditions of faulty communication are lack of communication and multiple meanings. However, [6] lists 12 different reasons for unnecessary costs, including lack

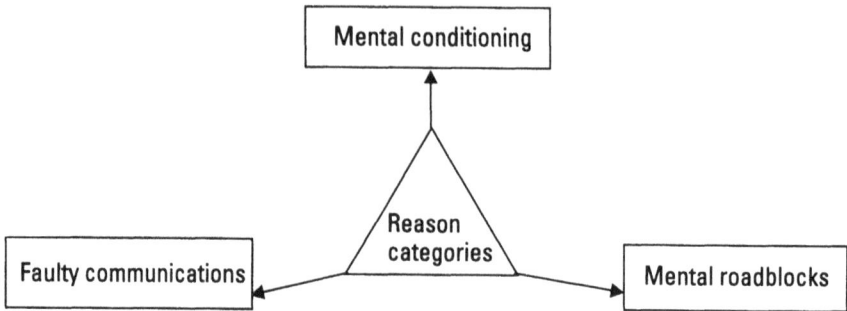

Figure 9.1 Basic categories of reasons for unnecessary costs.

of time, other responsibilities of engineers, lack of ideas, lack of information, lack of experience, desire to conform, preconceived ideas, prejudice, and failure to use available specialists.

Past experience indicates that organizations give reasons such as those listed here for not practicing value engineering [15].

- The organization is not big enough to practice value engineering.
- The organization's business is service oriented, and value engineering is for hardware.
- The company product varies widely in size, price range, quality, purpose, and use.
- The organization is involved in research and development, producing first-time, highly complex, and sophisticated products. The production units necessary to achieve the full potential of the value engineering approach are not there.
- The company purchases a large proportion of product parts; thus, the application of value engineering is not feasible.

9.4   Determining the Need for a Value Engineering Program

Prior to making the decision to install a value engineering program in an organization, it is important to conduct a thorough survey and study of its business, products, or services. This approach helps to assure maximum benefits from the application of value engineering. Answers to a list of statements presented in Table 9.1 can help to alleviate organization management's apprehension about the introduction of value engineering and decide whether or not to install it [15].

Table 9.1
Value Engineering Introduction-Related Statements

Number	Statement	True	False
1	The organization has a systematic approach to cost-effectiveness.		
2	Customers were asked about the "value" of company products.		
3	A small organization cannot afford to apply the principles of value engineering.		
4	The designers are not "creating" something that is already in existence.		
5	The company already has reliable cost data.		
6	The company is procuring as many standard components or parts as possible.		
7	The company has approached its specialty vendors for their assistance.		
8	The company designers thoroughly understand the true function of each and every part they design.		
9	The company personnel write effective reports.		
10	The company designers have opportunities to learn and practice creativity.		
11	It is not possible to adopt value engineering to company products because they are not produced in large volumes.		
12	The cost reduction programs currently employed assure a 10:1 return on investment.		
13	The company engineering professionals have sufficient time and ability to consider design-related economic factors.		
14	The company design engineers are kept up to date with the state of the art in technology and materials.		
15	The company business is process oriented; thus, the value engineering concept will not work in it.		
16	Any approach worth its salt will generate useful collateral benefits.		
17	It is not possible to try the value engineering approach on items or products of unusual technical complexity and sophistication.		
18	Value engineering is ineffective under situations where a high proportion of components for an item are procured.		
19	The company is a service-type business; thus, the concept of value engineering is not applicable.		
20	The company is too big; thus, the value engineering approach would be difficult and impractical to apply.		

Table 9.1 (continued)

21	Most of the company professionals have "satisfaction factors" built into their assigned positions.
22	The concept of value engineering is a panacea for all of an organization's economic-related problems.
23	The company cost programs normally assure collateral benefits, such as improved reliability and performance, for its products.
24	The application of the value engineering approach to company products is too risky because there is a high chance of lowering reliability and quality.
25	The consideration of the product cost in all company departments has been an important factor; thus, there is no need for value engineering.

For each correct answer to Table 9.1 statements, give four points. For most concerns, each Table 9.1 statement should be answered "false." If that is the case, the concern in question should expect a gratifying return from the investment in value engineering.

9.5   Value Engineering Phases

In the implementation of the value engineering concept, a systematic and multidisciplined approach is used. In the published literature, the phases of a value engineering job plan or approach may vary from 4 to 14, but it always contains essential core phases: information, speculation or creativity, and evaluation analysis [6]. These three phases cover the value analysis system's pivotal power and uniqueness.

For our purpose, we assume that the value engineering concept is composed of seven distinct phases, as shown in Figure 9.2 [5, 19]. Each of the Figure 9.2 phases will be discussed.

9.5.1   Team Selection Phase

This is concerned with the formation of the value engineering team. Careful consideration must be given in selecting team members, otherwise the inclusion of wrong members can be quite fatal to the value engineering effort. Furthermore, it is important to recognize that an individual's technical capabilities and experience alone may not be suitable for performing value engineering tasks.

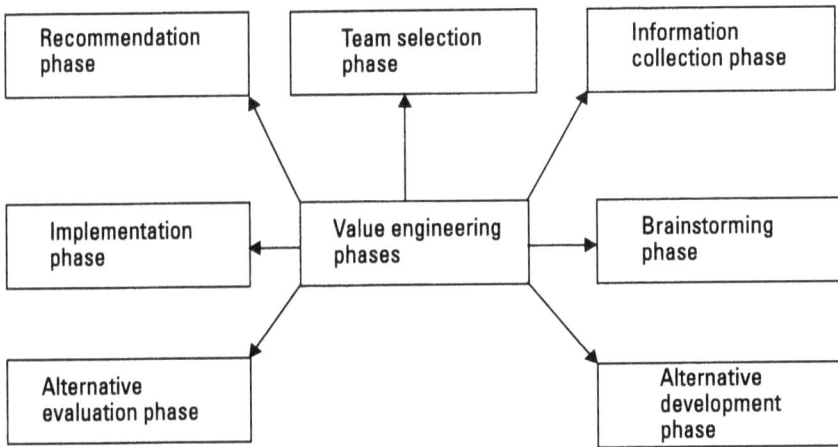

Figure 9.2 Value engineering concept phases.

With respect to the size of the value engineering team, [11] states that three to five persons are the minimum for efficient operation. In organizations of 3,000 or more employees, the value engineering staff is comprised of five to ten persons on the average [11, 20]. As per [12], usually a value engineering or management team is made up of three to seven individuals because more than seven members can lead to several drawbacks: Interaction becomes complex, the group begins to fracture, and discussions become indistinct. Also, an odd number is useful because it helps to reduce chances for split decisions [12].

Some of the important characteristics in forming a value engineering team are as follows [12, 21]:

- The team should be interdisciplinary and incorporate an effective balance of backgrounds, disciplines, viewpoints, and geographic representations. As the team must represent both product producer and user, it should include members from areas such as design, manufacturing, operations, marketing, and product planning.

- The team must include an expert on the product or subject under consideration.

- The team members should come from equivalent ranks within the organization so that peer pressure and politics are minimized.

- The team must include at least one member who is well versed in the value engineering concept.

- Past experience indicates that it is usually beneficial to have a specific decision maker on the team.

Some of the important factors concerning the value engineering team members are as follows [12, 22]:

- They are well aware of the sources of data in the areas of their expertise.

- They possess the characteristics to create, accept, and exploit change.

- They are at least familiar with the product or area under consideration.

- They possess an open mind and the ability to communicate and work with others in a team spirit.

- They possess the interest, motivation, and commitment to engage in the task under consideration.

- They will be able to obtain effective assistance and cooperation while representing their respective organizational sector.

- They will have an adequate amount of time to perform the job, and their length of association will be sufficient to provide continuity to the product.

9.5.2 Information Collection Phase

This phase is concerned with gathering technical and cost-related information from various sources. Some of these sources are vendors, clients, and in-house material from earlier similar projects. More specifically, internal sources for information include the system engineering department (i.e., for specification-related information), the manufacturing engineering department (i.e., for product manufacturing–related information), the accounting department (i.e., for detailed costs such as design, materials, production, labor, and overhead), and the purchasing department (i.e., for vendor-related information) [8]. For example, technical and cost-related information in environmental-related projects may be collected from such sources as the request for proposal, the contractor's proposal to the client, and design detail and process documentation.

Nonetheless, normally the types of information collected on a hardware project include engineering and cost information, procurement and manufacturing information, customer experience, reliability and quality-control information, tooling and testing information, and planning, scheduling, and shipping information [4]. By contrast, the types of information required for a nonhardware project include procedure flowchart, processing time, elapsed time, annual quantity processed, rate of usage, problems, delays, unsatisfactory features, signatures required, and history of previous changes [4].

Some of the useful guidelines concerning information collection are as follows [8].

- Verify information validity. For example, compare information from designers with engineering data books.
- Obtain information from the best sources.
- Compare with actual costs, including scrap, rework, and reinspection.
- Review scrap reports to determine the source of product problems.
- Determine the effectiveness in overcoming the roadblocks.

It is the responsibility of the value engineering team leader to collect information from individual team members and then present it to the whole team.

9.5.3 Brainstorming Phase

In this phase, various cost-saving alternatives are reviewed. More specifically, these alternatives are compared with the original design or function. Under such circumstances, the creativity factor plays a pivotal role in determining the most desirable alternatives. As the creative thinking of an individual value engineering team member is important to the success of a given project, careful consideration must be given to factors such as those listed next because they contribute to one's creative ability [8].

- *Originality.* This is the ability to develop unique solutions to a given problem.
- *Flexibility.* This is the ability to develop a wide range of methods or procedures to a given problem.
- *Sensitivity.* This is the ability to recognize an existing problem.

- *Fluency.* This is the ability to develop a substantial quantity of alternative solutions to a given problem.

There are three factors that inhibit the creative thinking of an individual: emotional, perceptual, and cultural. The emotional factors include fear of failure and desire for security; the perceptual factors include a failure to use all of the senses in observing, particularly the obvious; and the cultural factors include pressure from within and the surrounding environment to conform to already established patterns. Past experience indicates that the following guidelines are useful in developing a creative atmosphere [8]:

- Aim to keep the organization as flexible as possible so that emerging problems can be tackled with ease.

- Build an atmosphere that encourages the constant interchange of ideas and information.

- Encourage factors such as constructive nonconformity, diversity, individuality, and initiative.

- Provide leadership through persuasion rather than by orders.

- Recognize the accomplishments of individuals and reward them satisfactorily.

- Allow the maximum freedom possible at workstations.

- Permit as many involved professionals as possible to have an effective input in areas such as long-range planning and decision making.

- Provide diverse experience to individuals through courses, transfers, lectures, and attendance at meetings.

- Constantly seek out, develop, and encourage individuals with special talents.

In the past, many brainstorming methods have been developed. The group brainstorming method, whereby a group of individuals participate in an idea-generating session that lasts from 10 to 60 minutes, is probably the most widely used brainstorming technique [23]. This method is described in Chapter 7. Over the years, professionals working in the creativity field have developed many idea stimulators. These stimulators could be quite useful to individuals working in the value engineering area. Some of these stimulators are as follows [1]:

- Is it possible to change dimensions (i.e., larger, longer, smaller, shorter, thinner, thicker, deeper, or stand vertically)?

- Is it possible to change the cause or effect (i.e., stimulated, loader, altered, energized, strengthened, destroyed, counteracted, softer, or influenced)?

- Is it possible to adopt the use to a new market (i.e., children, men, women, foreigners, the handicapped, or the aged)?

- Is it possible to change the motion (i.e., deviated, oscillated, barred, animated, slowed, stilled, attracted, lowered, rotated, admitted, directed, speeded, repelled, agitated, or lifted)?

- Is it possible to change the quantity (i.e., more, less, join something, combine with something, fractionate, complete, or add something)?

- Is it possible to change the time element (i.e., slower, faster, shorter, longer, or synchronized)?

- Is it possible to change the state or condition (i.e., hotter, colder, softer, drier, wetter, lighter, heavier, or harder)?

- Is it possible to change the form (i.e., regular, irregular, rougher, smoother, curved, straight, or softer)?

- Can the character be changed (i.e., stronger, weaker, cheaper, substituted, reversed, add color, or change color)?

- Is it possible to change the order (i.e., arrangement, focus, assembly, or precedence)?

The factors listed in parentheses above are discussed in detail in [1].

9.5.4 Alternative Evaluation Phase

This phase is concerned with choosing alternatives with a high probability of providing cost savings and being implemented in the project design. Each of these alternatives is ranked with respect to its potential savings and likely acceptance of the client (if applicable).

9.5.5 Alternative Development Phase

This phase is concerned with studying in depth each chosen alternative to determine if it will yield the savings in cost initially anticipated. Furthermore, for each chosen alternative, a life cycle cost analysis, including capital, operating, and maintenance costs, is performed [24].

9.5.6   Recommendation Phase

This phase is concerned with presenting fully developed alternatives to the team by its leader.

The team chooses one or more recommendation(s) for reporting to the client (if applicable). Usually, these recommendations are presented in the form of a report. The report contains information on areas such as [6]:

- Study objective;
- Team members and areas of their specialties;
- Questionnaire data;
- Function diagram along with costs and customer attitude allocations;
- Value analysis targets;
- Proposal summary.

9.5.7   Implementation Phase

This phase is concerned with implementing the changes contained in the final report of the recommendation phase.

9.6   Application of Value Engineering to New Products— Related Factors and Value Engineering Techniques

From time to time, the application of value engineering to new products is considered. During that process, factors such as those listed next should be taken into account with care [8]:

- Size of production;
- Degree of tightness in design time;
- Similarity of new product to an old design;
- Profit margin;
- Competition intensity;
- Future business prospects;
- Significance of cost to customers;
- Value engineering required or not required by customers;
- Feasibility of changes;
- Type of existing performance and technical problems;

- Measurement of value engineering performance by customers;
- Severity of competition.

Over the years, a large number of value engineering techniques or guidelines have been identified, and their main objective is to help involved professionals to overcome various weaknesses. These techniques should be utilized as much as possible during value or engineering applications.

These techniques or guidelines include thinking creatively; using the value engineering plan; analyzing costs; using available resources; getting all of the important facts; working on specifics and avoiding generalities; using standards; getting new information; using good human relations; defining the function; blasting, creating, and then refining; identifying and overcoming roadblocks; evaluating the function; challenging requirements; spending the organization's money as you would your own; using information from only the best sources; and using your own judgment [4]. All of these techniques are described in detail in [12].

9.7 Management Responsibilities to the Value Engineering Program, Test for Value Questions, and Poor Value Factors

In any organization, management plays a key role in the success of a value engineering program. Furthermore, management participation is a primary prerequisite in achieving a total value engineering team effort, thus lower costs, higher profits, and greater enterprise growth. Some important management responsibilities to the value engineering program are as follows [11]:

- Providing adequate initial and continuing impetus to a value engineering program under consideration;
- Establishing global or overall value engineering objectives;
- Establishing a value engineering organization or department or group;
- Allocating a sufficient amount of budget to sustain the value engineering effort effectively;
- Providing an audit system on a continuous basis;
- Recognizing excellence in individual and group performance through appropriate means.

About half a century ago, one of the pioneer value engineers developed a test for value. This test is composed of a list of 10 questions. These questions can still be used today (i.e., with respect to parts or products) and are as follows [8]:

1. Does it require all of its current features?

2. Will anyone purchase it for less?

3. Does its application enhance value?

4. Is its cost proportional to its expected usefulness?

5. Is the item made on correct tooling, considering its quantity?

6. Is it possible that another reliable supplier can provide it for less?

7. Is there anything more suitable for the anticipated use?

8. Is it possible to find a standard item that can equally be usable?

9. Do anticipated profit, overhead costs, and labor and material costs add up to its cost, or is cost far greater than that?

10. Is it possible to manufacture a usable item by a lesser-cost approach?

Past experience indicates that many factors lead to poor value; these factors must be eliminated altogether. Some of these factors are temporary decisions, lack of essential information, nontroublesome items, nongeneration of ideas, predetermined reactions, personal inertia, and reluctance to seek advice [15]. All of these points are discussed in detail in [15].

9.8 Value Engineer's Functions and Characteristics and Value Engineering–Related Savings and Benefits

A value engineer, like other engineering professionals, performs various types of functions including investigation, fact collation, and, ultimately, practical application after installation. More specifically, some of the important tasks performed by a value engineer are risk identification, cost identification and reduction, motivation and indoctrination, preparation of program plans and reports, examination of part procurement, reviewing item simplification or elimination, and liaison with other groups [4, 5].

Some of the important characteristics of a good value engineer are the ability to be a good team player, good business judgment, adequate experience in cost analysis, substantial relevant technical experience, tact, flexibility, substantial relevant technical experience, skill in both oral and written communication, the ability to recognize different points of view, the ability to collect and organize information to identify potentially lucrative projects, self-confidence, and good emotional control [4, 5].

Over the years, value analysis or value engineering methodology has been applied across many sectors of economy and has generated substantial savings. For example, in 1989, the savings due to the application of value analysis/value engineering in various U.S. government departments have been presented in Table 9.2 [12].

Some of the benefits of practicing value engineering to a company are as follows [12]:

- Better quality at reduced cost;
- More efficient assembly;
- Better horizontal and vertical communications;
- Shorter ship-to-stock time;
- Shorter item-development time;
- Easier and simpler production or manufacturing methods;

Table 9.2

Approximate Savings in Various U.S. Government Departments, Due to the Application of Value Analysis/Value Engineering, for the Fiscal Year 1989

Number	Government Department	Approximate Savings in Millions of Dollars
1	Energy	59.4
2	Army	412.8
3	Agriculture	5.1
4	Navy	423.1
5	Air Force	488.6
6	NASA	0.8
7	Defense Logistic Agency	123.3
8	Interior	2.3
9	State Department	4
10	Transportation	12
11	General Services Agency	59.2
12	Veterans Administration	8.6

- Better value to item, product, customers, or users;
- More effective material selection;
- Lower number of postintroduction product or item problem solving;
- Better and more realistic specifications;
- Just-in-time kind inventories.

9.9 Problems

1. Write an essay on the history of value engineering.
2. Define the following terms:
   - Value engineering;
   - Value improvement;
   - Value analysis.
3. What are the principal objectives of value engineering?
4. What are the reasons for unnecessary costs?
5. Discuss the following phases of value engineering:
   - Team-selection phase;
   - Information-collection phase.
6. List at least 10 idea stimulators that could be useful to individuals working in the value engineering field.
7. Write down at least 15 value engineering techniques or useful guidelines.
8. What are the management responsibilities to the value engineering program?
9. Write down the test-for-value questions.
10. What are the characteristics of a good value engineer?
11. What are the benefits of practicing value engineering in a company?
12. Write down the important functions of a value engineer.

References

[1] Mudge, A. E., *Value Engineering: A Systematic Approach*, New York: McGraw-Hill, 1971.

[2] AMCP 706-104, *Engineering Design Handbook: Value Engineering*, Department of the Army, Washington, D.C., 1971.

[3] "Value Engineering Program," A STRICOM Document, Department of the Army, Washington, D.C., 2000.

[4] Heller, E. D., *Value Management: Value Engineering and Cost Reduction*, Reading, MA: Addison-Wesley, 1971.

[5] Dhillon, B. S., *Advanced Design Concepts for Engineers*, Lancaster, PA: Technomic Publishing Company, 1998.

[6] Fowler, T. C., *Value Analysis in Design*, New York: Van Nostrand Reinhold, 1990.

[7] Fallon, C., *Value Analysis*, New York: John Wiley and Sons, 1971.

[8] Brown, J., *Value Engineering*, New York: Industrial Press, 1992.

[9] Falcon, W. D., *Value Analysis-Value Engineering*, New York: American Management Association, 1964.

[10] Hales, J. A. G., *From Concepts to Capabilities*, New York: John Wiley and Sons, 1995.

[11] Greve, J. W., and F. W. Wilson (eds.), *Value Engineering in Manufacturing*, Englewood Cliffs, NJ: Prentice Hall, 1967.

[12] Shillito, M. L., and D. J. De Marle, *Value: Its Measurement, Design, and Management*, New York: John Wiley and Sons, 1992.

[13] Gage, W. L., *Value Analysis*, London: McGraw-Hill, 1967.

[14] Oughton, F., *Value Analysis and Value Engineering*, London: Sir Isaac Pitman and Sons Ltd., 1969.

[15] Clawson, R. H., *Value Engineering for Management*, New York: Auerbach Publishers, 1970.

[16] Ridge, W. J., *Value Analysis for Better Management*, New York: American Management Association, 1969.

[17] Miles, L. D., *Techniques of Value Analysis and Engineering*, New York: McGraw-Hill, 1972.

[18] Park, R., *Value Engineering*, Boca Raton, FL: St. Lucie Press, 1999.

[19] Acharya, P., and Z. C. Pfrommer, "Value Engineering," *Journal of Management in Engineering*, November/December, 1995, pp. 13–17.

[20] "Value Engineering Organization and Operation Report," *VE Bulletin No. 1*, Electronic Industries Association, New York, March 1963.

[21] Dillard, C. W., "Value Engineering Organization and Team Selection," *Proc. Society of American Value Engineers Conference*, 1975, pp. 11–12.

[22] Reigle, J., "Value Engineering: A Management Overview," *Value World*, Vol. 3, No. 3, 1979, pp. 4–8.

[23] Dhillon, B. S., *Engineering Management*, Lancaster, PA: Technomic Publishing Company, 1987.

[24] Dhillon, B. S., *Life Cycle Costing: Techniques, Models, and Applications*, New York: Gordon and Breach Science Publishers, 1989.

10

Reverse Engineering

10.1 Introduction

Reverse engineering may be seen as an unusual application of the art and science of engineering, but it has become a fact of daily life. It could be applied to rectify deficiencies in existing items or to extend their capabilities. In the early 1980s, it was used by the General Motors Corporation to maintain a competitive advantage over Ford Motor Company products (and vice versa). Furthermore, reverse engineering is used by major military powers on the equipment they can acquire from their antagonists. Sometimes, these military powers employ reverse engineering to provide spare parts and maintenance support to smaller powers that are no longer on friendly terms with the manufacturers of their existing weapon systems [1].

Broadly speaking, the business of reverse engineering does not differ greatly from that of detective work in a criminal investigation or from conducting intelligence operations for military. Currently, reverse engineering is practiced increasingly in industry at large because it offers many benefits, including reducing design and development costs and maintaining high-performance manufacturing capabilities [2]. It can also be used as an effective stopgap measure for improving system productivity until the resources required for full modernization are within reach. In general, it may be said that reverse engineering is directed at modernizing single-system elements, rather than total systems, for the purpose of maintaining or increasing system productivity.

This chapter presents different aspects of reverse engineering.

10.2 Reverse Engineering Terms and Definitions

This section presents the following selective terms and definitions directly or indirectly related to reverse engineering [1, 3, 4]:

- *Reverse engineering.* This is the process of developing a set of specifications of an item or a product by someone other than the designers of that item or a product, primarily based upon dimensioning and analyzing a specimen or a group of specimens.

- *Redocumentation.* This is the production of a semantically equivalent representation, frequently paper based, within the same relative abstraction level. Broadly speaking, redocumentation is the oldest and simplest form of reverse engineering.

- *Restructuring.* This is the transformation from one representation of a system into another without altering the subject system's external behavior (functionality and semantics).

- *Forward engineering.* This is the conventional process of moving from highly abstract and logical level, implementation-independent designs to the physical implementation of a given system.

- *Design recovery.* This may be described as the process whereby an approximation of the design of a subject item or system is achieved satisfactorily.

- *Resystemization.* This is the process of reverse engineering an item, eradicating environment-related design and implementation restrictions and subsequently forward engineering it subject to new environments without changing the functionality.

10.3 Reverse Engineering Objectives, Clone-Surrogate Reverse Engineering, and Basic Considerations

There are many different objectives of reverse engineering. In particular, with respect to software, some of them are to facilitate reuse, to recover lost information, to migrate from one hardware or software platform to another, to provide missing or alternative documentation, and to assist with maintenance [4].

When planning a reverse engineering project for the purpose of reproducing, say, a hardware item, it is necessary to determine whether the desired end result is to create a clone or a surrogate. In the case of creating a clone, reverse engineering means creating the identical or exact reproduction of the original (i.e., at least as far as circumstances will allow). More specifically, the clone reproduction must have the same form, operating mechanism, function, and fit as the original. By contrast, the surrogate item may carry the same function as the original in addition to being sized to fit in the same place as the original, but may neither appear to be the same, nor use the same operating mechanisms.

Obviously, the reverse engineering effort required in the case of the clone is much more extensive than that for the surrogate. The increased complexity and sophistication of modern equipment have made the task of producing clones even more difficult.

There are many basic considerations associated with the reverse engineering effort. The important ones are shown in Figure 10.1 [1].

The practice of reverse engineering requires two types of specifications (i.e., functional and dimensional), particularly for complex products. The functional specifications are useful because they describe the working of the

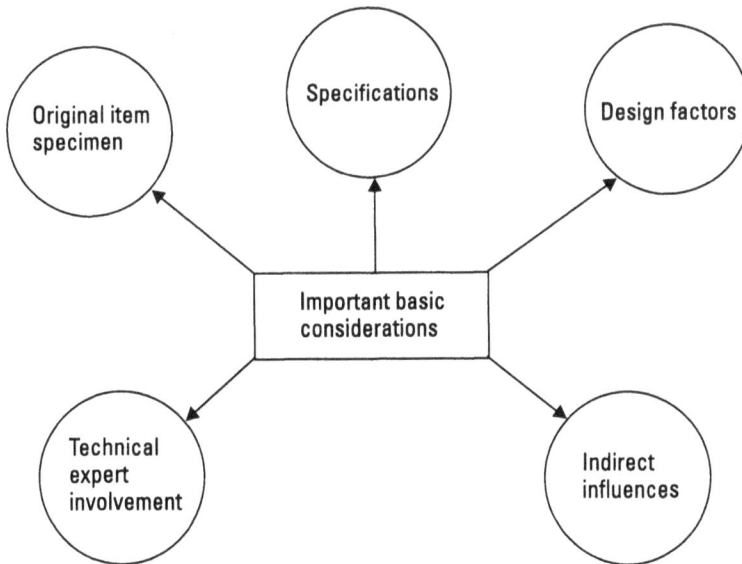

Figure 10.1 Important basic considerations associated with the reverse engineering effort.

product or item and its subsets, in addition their interactions. On the other hand, dimensional specifications include item dimensions (e.g., tolerances, lengths, and angles), materials to be used in part or item fabrication, how the parts or items are to be assembled and the materials are to be treated during manufacture, and parameter values and their associated tolerances.

With respect to design factors, it may be said that reverse engineering is significantly more cumbersome than executing an original design. Reasons behind this problem include the inability, in general, to determine the following factors: the thinking of the original designer, crucial parameters with respect to performance, treatments applied to the materials, and elements critical to the item's operation.

Indirect influences are normally one-time considerations during the reverse engineering effort associated with an item and include potential product users, maintenance policy, manufacturing philosophy, logistical support philosophy, and tactical deployment of item. However, giving careful consideration to indirect influences prior to undertaking reverse engineering can contribute immensely to its effectiveness.

The original item specimen is quite useful for making various types of decisions during the reverse engineering effort, such as testing hypotheses when everything else fails. Thus, it is absolutely essential to have at least one specimen of the intern or system under consideration for cloning in its original form.

With respect to technical-expert involvement, the reverse engineering effort normally acquires input from various different technical specialties. Therefore, it is often crucial to have input from relevant technical experts during the reverse engineering effort, as they can provide valuable information.

10.4 Reverse Engineering Method

The overall objective of this method is to provide a mechanism to direct work through all of its associated steps to the release of the specification, and to highlight clues to alert involved technical professionals that certain technical aspects of the work under consideration may not be that important to the cloning effort success. The following two assumptions are associated with this method [1]:

1. The subject item or system can be characterized as a hierarchical structure.

2. The method is applied repeatedly to the subject item until its reduction to piece parts or elements.

The grand plan for performing a reverse engineering effort is comprised of the following steps:

- *System engineer* to develop hypotheses based on available data and to highlight the measurement or test requirements. Also, assimilate existing data about the item to be reverse engineered, including its operation within the overall scheme of things.

- Disassemble to the level appropriate to verify or modify the hypotheses and to conduct supporting tests.

- Further system engineer based on all of the information in hand to develop new hypotheses as well as to prepare for additional measurement and testing.

- Disassemble further to measure and test to validate hypotheses and to uncover new information.

- Prepare specifications and documentation. This requires the engineer to continue the process until the level of understanding is adequate to do so.

10.5 Reverse Engineering Documentation

Because the findings of the reverse engineering process have to be communicated to all concerned individuals, it is imperative to adopt an appropriate documentation scheme. Furthermore, ensure that the documents used to record the findings of the reverse engineering process are compatible with the method used to guide the process. The following three types of documents are used in performing reverse engineering [1]:

1. *Equipment-breakdown hierarchy or the work-breakdown structure.* This is concerned with providing a mechanism to order the subsystems, their assemblies (including subassemblies), and parts of a subject system to expedite the development of specifications. The equipment-breakdown hierarchy document acts as a vehicle to guide the reverse engineering effort and is critical to the development of functional specifications. This document is also used to develop the configuration document.

2. *Configuration document.* This document is concerned with describing interconnections between various elements of a particular item, specifics of the flow of information, energy or materials between these elements, and the function(s) performed by them. The configuration document comprises several interrelated parts, including interface tables, block diagram(s), and functional description(s).

3. *Performance-specification document.* This is concerned with recording the performance specifications for the item by formulating a specification tree with the same structure as the equipment-breakdown hierarchy or structure. At all levels but the lowest (e.g., piece components), specification-tree entries describe the item's functional aspects. At the piece-component level, the entries in the performance specification are basically of the dimensional kind.

10.6   Software Reverse Engineering

As the software industry is growing at an alarming rate, the interest in software reverse engineering is increasing rapidly. For example, out of a total of 480 research articles and computing papers relating to or mentioning reverse engineering that appeared in 1990, more than 300 were specifically concerned with software reverse engineering [4].

10.6.1   Software Reverse Engineering Tasks and Tools

Software reverse engineering is comprised of a wide array of tasks concerning understanding and modification of software systems. Such tasks can be broken into a number of categories, including the four shown in Figure 10.2 [5].

In the case of Category I for some application domain, computer programs are representations of problem situations, and usually such programs do not contain any hint concerning the problem. The involved reverse engineer performs tasks such as reconstructing the mappings from the application domain to the program domain.

In the case of Category II, the software-development process follows from high-level abstraction to a rather detailed design and concrete implementation. The task of the reverse engineer is to move backward and develop an abstract representation of the implementation from the concrete detail mass.

The Category III tasks of software reverse engineering are concerned with detecting the objective and high-level structure of a program in case the

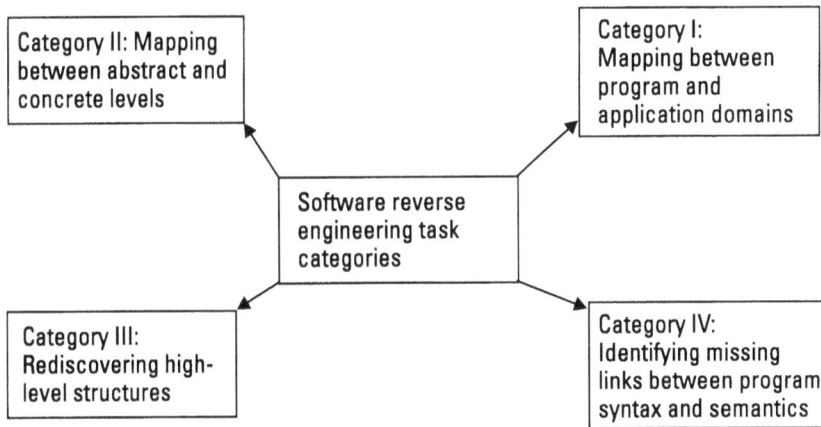

Figure 10.2 Some classifications of software reverse engineering tasks.

original one may have changed and in case no such single objective is left in the program.

In the case of Category IV, it may be said that computer programs are formal, particularly in the sense they have well-defined syntax and semantics. In addition, the meaning of a syntactically correct program determines the output for a particular input. Nonetheless, the systems requiring reverse engineering usually lose their original semantics. The software reverse engineering tasks determine the semantics of a specified program from its syntax.

Software reverse engineering makes use of various types of tools. Good tools can help to reduce cost and make software maintenance less troublesome. For example, if a reverse engineering tool can help to reduce maintenance cost by 5%, it would save billions of dollars [5].

Tools used in software reverse engineering projects may be classified into four major categories: software resource-analysis tools, code-converter tools, code-improvement tools, and true reverse engineering tools [4, 6, 7].

The software resource-analysis tools are marketed on the assumption that a programmer must spend a substantial amount of time determining which code to alter prior to actually altering a program code. These tools offer an effective mechanism for developing comprehensive module/job step/program/job cross-references. Their more advanced capabilities include the assessment of program complexity with respect to the processing they expect to achieve and of the degree to which an individual program can be considered well structured.

In reverse engineering terms the tools belonging to this category may be technologically the simplest, but they are usually quite appealing to information systems people in the short term.

The code-converter tools are used to convert code from one language to another, (e.g., C to COBOL). Often, they could be quite complex and sophisticated, even emulating source-support environments on the objective of target platform. The simple reason for their use in reverse engineering exercises is to make more tools available. Even with the best converters, code conversion is not a simple and straightforward matter, and in many cases, it would be a better approach to adapt a chosen reverse engineering tool by commissioning a new "front-end" parser.

The code-improvement tools include restructurers, reformatters, debuggers, and data standardizers. The basic objective of using all of these tools is to improve one way or the other the effectiveness of reverse engineering.

Reverse engineering tools reverse code and/or data by one or more levels of abstraction, populating a repository with the end artifacts and all of their possible relationships. Available tools vary with respect to abstraction, level of automation, and the richness of the repository. All in all, there are more than 40 true reverse engineering tools that can operate over a wide range of hardware and software platforms [4].

Prior to selecting tools for use during reverse engineering, it is important to clearly identify the overall and intermediate objectives. Nonetheless, in selecting such tools, one should ask questions on areas such as language coverage, techniques supported, type and level of abstraction, application environment, tailoring, reference sites, industry standards, integration, repository support, and vendor support.

10.6.2 Software Reverse Engineering Application Areas and Project Selection

There are many areas in which software reverse engineering is applicable. Most of these areas are shown in Figure 10.3 [5].

However, the processes that attempt to redesign, restructure, and enhance functionality of a system fall outside the scope of reverse engineering.

The systems that can be considered appropriate candidates for reverse engineering exhibit characteristics such as poorly structured code, outdated documentation, incomplete or missing design specifications, overly complex modules, requirement for migration to a new software platform, need for excessive corrective maintenance, hard-coded parameters subject to change, and need for migration to a new generation of hardware.

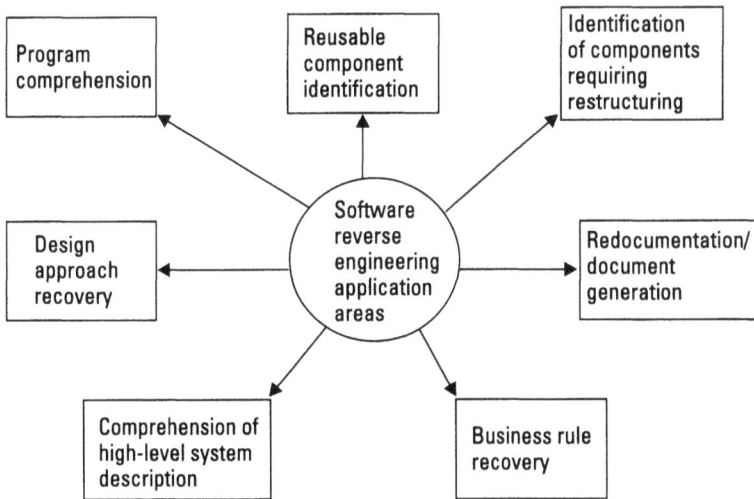

Figure 10.3 Software reverse engineering application areas.

In selecting a suitable system for reverse engineering, questions such as these should be asked [4]:

- Is the code in question well structured?
- Is the system under consideration crucial to the operation of the enterprise?
- Is the underlying business logic highly dynamic?
- Are reverse engineering tools within reach for the environment supported?
- To what degree does the system under consideration interface with other systems?
- How many languages does the system comprise?
- To what degree are the in-house subroutines used in the system?
- How will the reverse engineering impact current working practices?
- Are the original professionals, such as analysts, designers, and programmers, still working for the organization?
- Are any of the structured techniques currently being used in development projects?
- Does the involved staff clearly understand forward engineering methods or techniques?

10.6.3   Software Reverse Engineering Benefits, Difficulties, and Related Pointers

There are many benefits of software reverse engineering. A reverse engineering project should be expected to deliver benefits in areas such as shown in Figure 10.4 [4].

Some of the difficulties associated with the practice of software reverse engineering are as follows [8]:

- *Volume of code.* A very large code base requires analysis by some method or technique.

- *Lack of proper documentation.* Usually, system documentation is out of date and in many cases nonexistent.

- *Lack of proper knowledge or expertise.*

- *Systems are not streamlined.*

- *Inconsistent standards.* Systems do not correspond to one set of standards.

Over the years people working in the area of software reverse engineering have developed many pointers for the benefit of people involved with the practice of reverse engineering. Some of the important ones are as follows [4]:

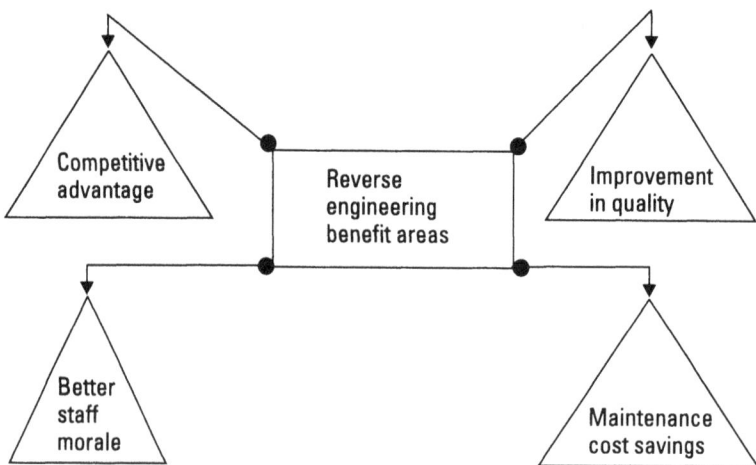

Figure 10.4 Areas of software reverse engineering benefits.

- Complete automation is impossible.
- Staff motivation and training are a crucial factor.
- On the average, large systems have three different languages (e.g., JCL, COBOL, and Assembler).
- Most installations have nonstandard language variants.
- Data transformation or conversion may be needed.
- Be extra cautious about reverse engineering into a *black hole* or cul-de-sac.

10.7 Reverse Engineering–Related Court Cases and Laws

Over the years, there have been various court decisions on cases regarding whether copying programs for reverse engineering constitutes copyright infringement [9, 10]. Some of the cases involved are presented next.

Case I.   This is often referred to as the *NEC v. Intel* copyright case, and it involved reverse engineering. In February 1989, this case was decided in a federal court in California. NEC admitted disassembling the Intel 80861 8088 microcode, and Intel heavily relied on similarities in the microcode subsequently developed by NEC (rather than the disassembling) for its copyright-infringement claim. The factor that played an instrumental role in persuading the presiding judge that the similarities in both microcodes stemmed from constraints due to hardware, architecture, and specifications, and thus did not infringe the Intel copyright, was that NEC had set up clean-room process to separate the analysis of the Intel microcode from the development of NEC's functionally equivalent microcode.

Case II.   This is known as the *MAI v. Hubco* case. In this case, MAI sued Hubco on two points: infringement of its copyright of operating-system software and misappropriation of the trade secrets in the software. The court's decision stated that Hubco did not steal any trade secrets from MAI because it deciphered them through reverse engineering. However, the court did issue a preliminary injunction against Hubco's use of its software program, in addition to its on-site visits to MAI customers to remove the governors.

Case III.   This is known as the *SAS Institute v. S&H Computer Systems* case, involving the copying of a program for reverse-engineering purposes. SAS was successful in charging the developer (i.e., S&H Computer Systems) of a competitive statistical-analysis program with respect to copyright infringement and trade-secret misappropriation. In this case, S&H acquired a copy of the source and object code of the SAS software through a licensing agreement for the purpose of studying it so that S&H could produce a similar type of program that could run on a different kind of computer system than the one on which SAS ran.

Case IV.   This is known as the *E. F. Johnson v. Uniden* case, involving the development of a competitive software program in part by disassembling the plaintiff's programs. The court ruled that disassembly alone was not copyright infringement. Nonetheless, in the final judgment, the court ruled that Uniden infringed on the copyright law on ground that the two programs in question were more similar in the descriptions of their implementation than was describable by the requirement to be compatible with the Johnson radio-system software.

There are many national and international laws that are directly or indirectly concerned with reverse engineering, including trade-secret law, semiconductor-chip law, patent law, copyright law, Japanese copyright law, and British copyright law [9, 11–20]. Two of these laws are discussed next [9].

Trade-secret law.   This is said to be the first U.S. intellectual-property law to establish a rule concerning the legality of reverse engineering. Trade-secret owners have to be protected against misappropriation of their trade secrets. There are two ways to misappropriate trade secrets: (1) the owners disclosed the secret in confidence to another, and (2) application of improper means to acquire the trade secret (e.g., bribery or industrial espionage). Trade-secret law provides the right to be free from unwarranted interference with confidential relationships and use of improper channels to obtain a secret, but it has traditionally not protected any property interest as such in a trade secret. Consequently, anyone can learn another's trade secret through reverse engineering, thus acquiring the trade secret legally. Thus, under the trade-secret law, reverse engineering is considered to be a good thing, even though it may possess the potential to undermine some economic interests of trade-secret owners.

U.S. semiconductor-chip law.   This contains a special reverse engineering provision that allows competitors to copy a protected mask work for studying the concepts/techniques embodied in the mask work or in the chip's logic flow, circuitry, or organization. Also, the law permits the incorporation of the results of this effort in a competitive chip item made up of original work that is not substantially identical to the protected mask work studied.

10.8 Reverse Engineering Team and Tips for Success

The performance of reverse engineering is not a one-person task. It requires a group of specialists, including estimators, drafts people, technicians, shop personnel, engineers, and production workers [21]. From time to time, specialists in such areas as metallurgy, ceramics, circuit design, and vibration analysis are also needed. However, normally only a small number of individuals form the core reverse engineering team.

Nevertheless, it is generally important to retain the same core team members from project to project to maintain consistency and enable them to build on their experience. The individual chosen to lead the team should be a generalist with some knowledge of engineering disciplines in such areas as electrical and mechanical engineering, manufacturing, and electronics. Because the team leader will need to interact with various people, he or she should also have good managerial abilities.

The selection of items for the application of reverse engineering requires careful consideration. Normally, good reverse engineering candidates have such characteristics as excessive cost, high usage, and a high failure rate. To determine whether an item will make a suitable candidate, one should consider such issues as potential return on investment, economics, technical complexity and criticality, and logistics [2, 22]. Additional factors to consider include patent rights for the item in question (e.g., who owns them?), the adequacy and availability of technical data on the item, support obsolescence (e.g., support items are out of date), and lack of supply of required parts.

To ensure that the reverse engineering effort is successful, one should:

- Expect that a good reverse engineering program will take from 2 to 5 years to become self-sufficient.

- Generally, aim for a return on investment on prescreened candidates of at least 25:1. The expected return on investment for high-risk projects should be at least 200:1.

- Aim for a minimum 25% reduction in the item's unit cost.

- Make only a moderate investment during the initial stage of the program to avoid spending too much money without some certainty of success.

10.9   Traditional and Reverse Engineering Design Processes and Future Reverse Engineering Challenges

Traditional and reverse engineering design processes differ quite significantly. The stages of a traditional design process can vary from a minimum of four to a maximum of 25 [22]. For example, the minimum four stages are requirement, design idea, prototype and test, and product. By contrast, usually the reverse engineering design process comprises six stages: product, disassembly, measure and test, design recovery, prototype and test, and reverse engineered product [21].

Reverse engineering is faced with many future challenges. In particular, with respect to software, some of the challenging areas are as follows [23, 24]:

- Test and validation of reverse engineering technology;
- Clearer articulation of the reverse engineering process;
- Prevention of software systems, being created currently, from becoming the incomprehensible legacy systems of tomorrow;
- Role of management in the success of reverse engineering technology;
- Raising the conceptual level at which software tools interact and communicate with people such as domain experts, users, and software engineers;
- Measurement of potential impact of reverse engineering.

10.10   Problems

1. Define the following terms:
   - Reverse engineering;
   - Restructuring;
   - Redocumentation.
2. Discuss basic considerations associated with the reverse engineering effort.
3. Describe the grand plan for performing a reverse engineering effort.
4. Discuss three types of documents used in performing reverse engineering.
5. What are the four major classifications of software reverse engineering tasks?
6. List at least seven software reverse engineering application areas.

7. List at least 10 questions useful for selecting a suitable system for reverse engineering with respect to software.

8. What are the difficulties associated with the practice of software reverse engineering?

9. Discuss two reverse engineering–related court cases.

10. Discuss the membership of the reverse engineering team.

11. What are the important benefits of reverse engineering?

References

[1] Rekoff, M. G., "On Reverse Engineering," *IEEE Trans. on Systems, Man, and Cybernetics,* Vol. 15, No. 2, 1985, pp. 244–252.

[2] Dhillon, B. S., "Reverse Engineering: A Design Approach Whose Time Has Come," *Engineering Dimensions,* September/October 1998, pp. 19–20.

[3] Chikofsky, E. J., and J. H. Cross, "Reverse Engineering and Design Recovery: A Taxonomy," *IEEE Software,* January 1990, pp. 13–17.

[4] Frazer, A., "Reverse Engineering: Hype, Hope, or Here," in *Software Reuse and Reverse Engineering in Practice,* P. A. V. Hall (ed.), London: Chapman and Hall, 1992, pp. 210–243.

[5] Ashrafuzzaman, M., "Reverse Engineering," *Computer Based Learning Unit,* University of Leeds, Leeds, U.K., 1995.

[6] Holloway, S., "Re-engineering Business Systems to Use the Next Generation of Software," in *Software Reuse and Reverse Engineering in Practice,* P. A. V. Hall (ed.), London: Chapman and Hall, 1992, pp. 271–282.

[7] Jones, R., "A Review of Current CASE Technology," in *The Distributed Development Environment: The Art of Using CASE,* S. R. Holloway (ed.), London: Chapman and Hall, 1990, pp. 100–120.

[8] McGill, R., "Reverse Engineering: Not Yet?" in *Software Reuse and Reverse Engineering in Practice,* P. A. V. Hall (ed.), London: Chapman and Hall, 1992, pp. 245–252.

[9] Samuelson, P., "Reverse Engineering Someone Else's Software: Is It Legal?" *IEEE Software,* January 1990, pp. 90–96.

[10] Laurie, R. S., and S. M. Everett, "Protection of Trade Secrets in Object-Form Software: The Case for Reverse Engineering," *Computer Lawyer,* July 1984, pp. 1–11.

[11] Samuelson, P., "Contu Revisited: The Case Against Copyright Protection for Programs in Machine-Readable Form," *Duke Law J.,* October 1984, pp. 663–752.

[12] Samuelson, P., "Modifying Copyrighted Software: Adjusting Copyright Doctrine to Accommodate a Technology," *Jurimetrics J.,* Winter 1988, pp. 179–221.

[13] Mishra, R., "Reverse Engineering in Japan and the Global Trend Towards Interoperability," *Murdoch University Electronic Journal of Law*, Vol. 4, No. 2, 1997, pp. 1–12.

[14] Dworkin, G., "The Concept of Reverse Engineering In Intellectual Property Law and Its Application to Computer Programs," *The Intellectual Property Law Journal*, Vol. 1, 1990, pp. 164–165.

[15] Karjala, D., "The First Case on Operating Systems and Reverse Engineering of Programs in Japan," *EIPR*, Vol. 6, 1988, pp. 172–175.

[16] Doi, T., "Computer Technology and Copyright: Legislative and Judicial Developments," *Michigan Yearbook of International Legal Studies*, Vol. 8, 1987, pp. 3–24.

[17] Nakajima, T., "Legal Protection of Computer Programs in Japan: The Conflict Between Economic and Artistic Goals," *Columbia Journal of Transitional Law*, Vol. 27, 1988, pp. 148–156.

[18] Clapes, A. L., *Softwars: The Legal Battles for Control of the Global Software Industry*, London: Quoron Books, 1993.

[19] Karjala, D., "Lessons from the Computer Software Debate in Japan," *Arizona State Law Journal*, 1984, pp. 55–65.

[20] Durney, E. G., "Protection on Computer Programs Under Japanese Copyright Law," *UCLA Basin Law Journal*, Vol. 9, 1991, pp. 42–50.

[21] Ingle, K. A., *Reverse Engineering*, New York: McGraw-Hill, 1994.

[22] Dhillon, B. S., *Engineering Design: A Modern Approach*, Chicago, IL: Irwin, 1996.

[23] Wills, L. M., and J. H. Cross, "Recent Trends and Open Issues in Reverse Engineering," *Automated Software Engineering*, Vol. 3, 1996, pp. 165–172.

[24] Selfridge, P. R. W., and E. Chikofsky, "Challenges to the Field of Reverse Engineering," *Proc. First Working Conference on Reverse Engineering*, May 1993, pp. 144–150.

11

Configuration Management

11.1 Introduction

Configuration simply refers to the arrangement of the parts or elements of something, and management refers to the act or practice of managing. By combining both words, it could be stated that configuration management is simply the act of managing the arrangement of parts of an item, product, or system. References [1–3] define configuration management as a discipline for providing a systematic mechanism for planning, identifying, controlling, and accounting for the status of an item or system's configuration, from its birth to death.

The history of configuration management may be traced back to the 1950s when the professionals involved with the missile launch program found that prototype units were in orbit and there was poor documentation of changes made to the unit in question, specifically part number identification records [4]. In 1962, the U.S. Air Force released a document entitled "Configuration Management During the Development and Acquisition Phases" (AFSCM 375-1) and revised it 2 years later [5]. Also in 1964, the National Aeronautics and Space Administration (NASA) released a similar document entitled "Apollo Configuration Management Manual" (NPC 500-1) [6]. In 1965, the U.S. Army materiel command released AMCR 11-26, another document on configuration management. In 1967, the U.S. Navy issued its own document on configuration management (i.e., NAVMATINST 4130.1).

In the latter part of the 1960s, U.S. Secretary of Defense Robert B. Mac-Namara issued a directive calling for a single system of configuration management for all three services [1]. Consequently, in 1968, the Department of Defense released a document titled "Configuration Management" (DOD Directive 5010.18) [1]. Subsequently, the Department of Defense prepared the following military standards concerning configuration management [1, 6]:

- MIL-STD-483, Configuration Management Practices for Systems, Equipment, Munitions, and Computer Programs, 1970;

- MIL-STD-482, Configuration Status Accounting Data Elements and Related Features, 1970;

- MIL-STD-1456, Contractor Configuration Management Plans, 1972;

- MIL-STD-481A, Configuration Control-Engineering Changes, Deviations, and Waivers (short form), 1972.

In 1992, the U.S. Department of Defense issued another standard titled "Configuration Management" (MIL-STD-973) that superseded the existing military standards: 480-483, 1456, and 1521 [1]. In 1998, the American National Standards Institute released a standard titled "National Consensus Standard for Configuration Management" (ANSI/EIA-649-1998) [7].

Over the years many professionals have contributed to the subject of configuration management and this chapter presents different aspects of this topic.

11.2 Configuration Management Terms and Definitions

This section presents selective terms and definitions directly or indirectly associated with configuration management [1, 5, 7–11].

- *Configuration management.* This is a discipline for providing a systematic mechanism to plan, identify, control, and account for the status of the configuration of an item or system, from its birth to death.

- *Software configuration management.* This is configuration management tailored specifically to systems, or parts of systems, that are predominantly made up of software.

- *Configuration management plan.* This is a document that describes the detailed requirements for identification, control, status accounting, and audits necessary to manage the configuration of a specified item.

- *Configuration planning.* This is the process of determining how an item or product shall be configured, supported, and documented.

- *Configuration item.* This is any part of the development and/or deliverable system that is required to be independently identified, tested, used, reviewed, stored, delivered, changed, and/or maintained.

- *Configuration control.* This is a systematic approach to identify, control, and account for the status of changes that affect the configuration or documentation of an item.

- *Baseline.* This is an approved reference point, at a specific point in time, to control an item's design, development, manufacture, and maintenance.

- *Configuration identification.* This is the technical documentation that is appropriately identified and defines the accepted configuration of items under design, development, and test, in production or in the field.

- *Engineering change request.* This is a form that is available to any individual in an organization to use when trying to provide a detail of a proposed change or problem that may be present on a given item or product.

- *Configuration status accounting.* This is the process of knowing the documentation status of each item or product by serial number before its shipment to a customer and knowing its status in the field.

- *Configuration control board.* This is a group of technical and administrative professionals responsible for reviewing engineering changes made to the configuration item after the approval of the baseline.

- *Configuration audit.* This is the verification, through inspection, that established standards and procedures are implemented and followed effectively by the project people responsible, and that the configuration as per the status reports clearly reflects that actually being installed, tested, or used.

11.3   Types of Product Changes and Their Reasons During Design and Production

There are basically three types of changes made to a product or item [6]:

1. Changes that result in significant savings in a product's life cycle cost;

2. Changes that are needed to rectify deficiencies in the item or product;

3. Changes that are useful in improving an item or product's operational use or its logistic support.

Some of the reasons for making changes during the design and development phases of a product are listed here [6].

- Weaknesses are found in the proposed design of the item through analysis, testing, or another phase.

- The designer of the item or product finds new ways to meet the same objective but more efficiently.

- Production personnel experience difficulties in producing the item on time as per the original design.

- Better parts or components become available due to advancement in the state of the art.

- The production department discovers a way to enhance item or product performance. Usually this is quite rare.

- The production department substitutes components or materials to reduce cost.

11.4   Reasons for and Objectives of Having a Configuration Management System

There are many reasons for having a configuration management system and some of these are listed here [1].

- It provides a systematic approach for a good engineering documentation control system.

- The use of both hardware and software in modern items or products needs an effective system of configuration management.
- Complex customer and manufacturer or contractor interrelationships now exist due to the technical complexity of the modern products.
- It is an approach that provides all the elements required for an effective engineering documentation control system.
- ISO 9000 Series of standards on quality systems call for having an effective configuration management system to control new items or products.
- The increasing frequency of product liability suits against companies makes it necessary to have a good configuration management system.

Some of the main objectives of having a configuration management system are as follows [1, 5, 12, 13]:

- To plan, identify, control, and account for the status of an item's configuration;
- To have an efficient change-control system;
- To have uniform application of all procedures associated with an effective configuration management system by all involved individuals within the organization;
- To avoid stifling the creative capabilities of technical professionals;
- To achieve logistic support of an item at the minimum life cycle cost.

11.5 Configuration Management Plan and Disciplines

The configuration management plan divides the product/item life cycle into four basic phases, as shown in Figure 11.1. These are: concept formulation phase, definition phase, acquisition phase (i.e., design and development stage and production stage), and operational phase [5, 14]. As shown in Figure 11.1, three baselines terminate the first three phases: characteristics, the concept formulation phase; functional, the definition phase; and operational, the acquisition phase.

Furthermore, the product or item sub-baseline denotes the termination of the design and development stage. All of these baselines represent checkpoints during the item or product life cycle. They simply mean that a new

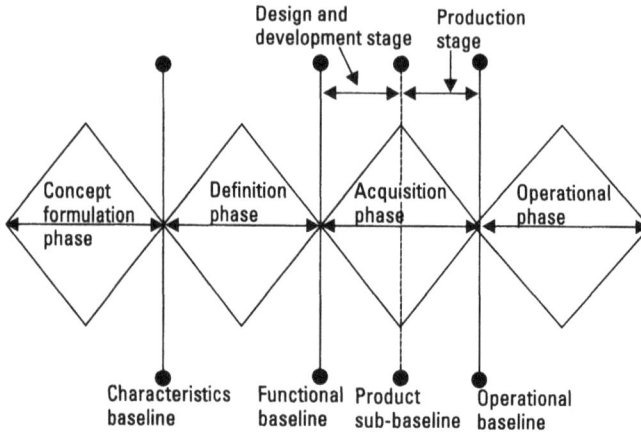

Figure 11.1   Configuration management plan phases and baselines.

phase cannot be initiated unless all of the details and issues raised during the preceding phase are resolved in a satisfactory manner.

During the product or item life cycle, the configuration management disciplines of identification, control, and accounting are used.

11.5.1 Identification

This plays a pivotal role during the early phases because it is during the concept formulation phase, and, to a lesser degree, the definition phase, that the objectives, ideas, and concepts are formulated. The identification discipline calls for documenting complete and accurate descriptions of the product during each life cycle phase. More specifically, the configuration identification may be described as the technical documentation that is properly identified, and it defines the approved configuration of products or items under design, development, test, production, or in the field. Some of the documentation required under identification includes product specifications, drawings, the bill of material, assembly parts lists, spare parts lists, manuals, and software documentation. Examples of the definition-phase documentation items are specification, contract, and schedule. The important requirements of a good configuration identification system are as follows [1, 5]:

- It is simple to develop, understand, and use.
- All associated data and equipment items are identified.
- An effective system is developed with written procedures to control identification operations.

- It is contemporary so that the state-of-the-art technology of a new product design can be handled effectively.

- It caters to an individual or a group to be responsible for issuance and control of all associated configuration identifiers.

- An independent control group of qualified individuals is established to ensure that proper identification procedures are being practiced.

- It is quite flexible so that it can be adapted to different situations without any difficulty.

11.5.2 Control

This is a continuous function that starts in the initial stages of a project and extends over the entire service life of the individual configuration item. Configuration control may be described as a systematic or organized approach to identify, control, and account for the status of changes that affect the item or product configuration or documentation. Some of the configuration control–related policies and objectives are as follows [5]:

- All data (i.e., engineering and configuration) affected by formally approved changes will be revised as considered appropriate to describe the change.

- All approved changes to engineering configuration data will be effectively implemented so that the cost associated with the changes is minimal and the configuration item is produced as specified.

- Only changes to areas such as meeting technical requirements, reducing cost, improving product or item performance, or providing benefits to customers will be proposed and implemented.

- All actions concerning the submission, analysis, approval, and implementation of changes will be documented effectively.

- Configuration management disciplines will be developed and maintained on all items subcontracted.

- Formally approved changes will lead to appropriate corresponding changes to all related support materials including handbooks, support equipment, and manuals.

- Retrofit engineering changes will only be recommended for incorporation when mandatory for functional reasons or when retrofitting is beneficial to the customer.

Prior to the acceptance of a product or item by the buyer or customer, the configuration control effort is primarily directed on controlling the configuration as per specification or other similar documents. However, subsequent to the acceptance of a product or item by the customer or buyer, the configuration control is focused basically on hardware with changes to documentation occurring because of approved specification and hardware changes.

Change Analysis Factors

After the approval of a configuration manager, the configuration administration and planning professionals perform analysis of the change with respect to factors such as follows [5]:

- Number of parts due from vendors;
- Number of parts being fabricated in-house and their stage of completion;
- Test qualification need and qualification status;
- Models and projects in which the parts are used;
- Number of parts shipped and in spare inventory;
- Number of parts received from vendors and their location.

11.5.3 Configuration Status Accounting

This establishes records that enable the establishment of appropriate logistics support. More specifically, configuration status accounting may be described as the process of knowing each item's documentation status by serial number prior to its shipment and knowing its status in the field environment. The records include items such as those shown in Figure 11.2.

Configuration Accounting Report

A configuration accounting report includes various types of information [5]: reason for the change, urgency of the change, identification date of the change, who identified the change, estimated cost of the change, actual cost of the change, change description, scheduled submittal date, actual submittal date, technical manual required, technical manual to be changed, required and actual decision dates, required and actual authorization dates, the decision made, authorization document, scheduled and actual engineering release, scheduled and actual procurement, receipt of material, scheduled tool fabrication, actual tool fabrication, place of installation, who is responsible for

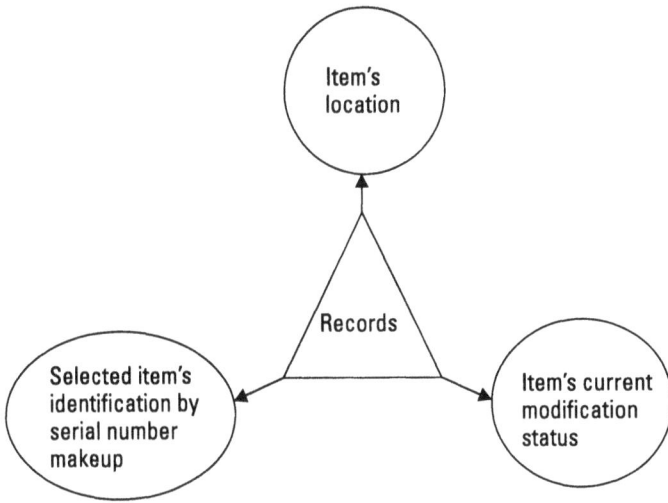

Figure 11.2 Record items.

installation, man-hours needed for installation, system downtime during installation, scheduled and actual part fabrication, scheduled and actual assembly, scheduled and actual installation, spare parts to be reworked, and new spare parts required.

11.6 Configuration Management Organization

The success of configuration management depends on the effectiveness of the configuration management organization or department. In this aspect, one of the most crucial factors is the location of the configuration management department within the structure of the engineering organization. Although the configuration management chief may report directly to the executive of engineering, in large companies the engineering department usually has an organization known as engineering services, and it is within the framework of this organization that the configuration management organization is normally located.

Some of the people who may belong to the configuration management organization are as follows:

- Data entry clerk;
- Drafter;

- Technical writer;
- Secretary;
- Designer.

The configuration administration and planning staff support the manager of the configuration management organization, and some of its main responsibilities are as follows [5]:

- Process effectively to completion all engineering change–related requests started on the project.
- Liaise with customers as appropriate.
- Coordinate activities concerning configuration control board.
- Perform the implementation planning for all changes.

11.6.1 Configuration Control Board

Some organizations establish a configuration control board for presenting data and applicable arguments concerning proposed changes by members representing all important functions. Within the framework of a project, the board is the final authority on all major proposed changes. As the configuration control board is not a voting board, the board chairman has the absolute authority to make decisions. Past experience indicates that the configuration control board could be a very valuable group, if it meets regularly.

The board members belong to areas such as these:

- Systems engineering;
- Contracts;
- Manufacturing;
- Logistics;
- Material or purchasing;
- Design and development;
- Integration and test;
- Project control;
- Product integrity.

The principal function of the board members is to verify items such as necessity for change, the meeting of schedule and cost requirements, and the feasibility of the implementation method.

The board members consider many factors, as presented in Table 11.1, with respect to the proposed change and its effects on other project areas.

Table 11.1
Factors Considered by the Board Members with Respect to the Proposed Change and Its Effects on Other Project Areas

Number	Factor
1	Impact on delivery schedule and project cost
2	Total configuration item's performance
3	Configuration item's reliability
4	Implementation by manufacturing
5	Configuration item's service life
6	Configuration item's safety
7	Part procurement problems
8	Special tooling need and cost
9	Rework on previously delivered configuration item
10	Difficulty in repair and maintenance
11	Configuration item's electrical and mechanical installation
12	Scrapping of completed assemblies without the changes
13	Spare parts
14	Configuration item's weight, size, stability, and power consumption
15	Changes in manuals, drawings, specifications, and procedures
16	Impact on the magnetic and radio interference characteristics of the configuration item
17	Modification of the qualified configuration item
18	Safety of operator or others
19	Need for additional testing
20	Impact of the additional testing on schedule and cost
21	Efficacy of the change incorporation

11.7   Configuration Manager Responsibilities and Qualities

Just like any other engineering manager, the configuration manager also has various responsibilities. His or her responsibilities may be grouped into three main areas as shown in Figure 11.3. In order to perform these responsibilities effectively, the configuration manager must possess qualities such as follows [5]:

- Good documentation skills;

- Ability to administer and organize;

- Attention to details;

- Tact and diplomacy;

- Good technical knowledge in areas such as design, manufacturing, testing, and quality control;

- Familiarity with the company and the product;

- Ability to coordinate the work of individuals in widely different activities;

- Ability to focus on the overall picture of a project and at the same time not ignore minute details;

- Ability to resist pressure to take shortcuts during an emergency.

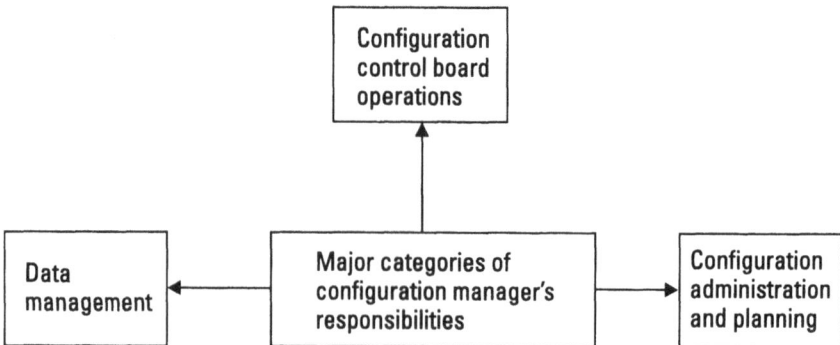

Figure 11.3 Main categories of a configuration manager's responsibilities.

11.8 Successful Configuration Management Characteristics and Guidelines for Controlling Products' Configuration and Documentation

Just as in the case of any administration and management field, certain characteristics are important to the success of configuration management. Globally, it may be said that a successful configuration management activity satisfies its assigned goals and objectives at minimum cost to the buyer and seller. The important features that characterize successful configuration control, identification, and accounting are listed next [5]:

- Accurate, complete, and early definition of configuration management scope, procedures, goals, and objectives;
- Simple configuration control, identification, and accounting approach or system—as much as possible;
- Minimum need for labor;
- Accurate and complete descriptions of changes in question;
- Efficient change evaluation and processing;
- Minimal number of forms and other associated documents to implement changes and to provide full records of such changes;
- Cooperative and responsive customers;
- Effective coordination among project team members;
- Accuracy in change identification and accounting.

Some of the guidelines for controlling products' configuration and documentation are shown in Figure 11.4 [1].

11.9 Symptoms of Ineffective Application of Configuration Management

Usually, a company involved in equipment manufacture practices some form of configuration management. Some of the important symptoms of ineffective application of the configuration management approach to a given project are as follows [14]:

- Unclear objectives;
- Excessive time to make change related decisions;

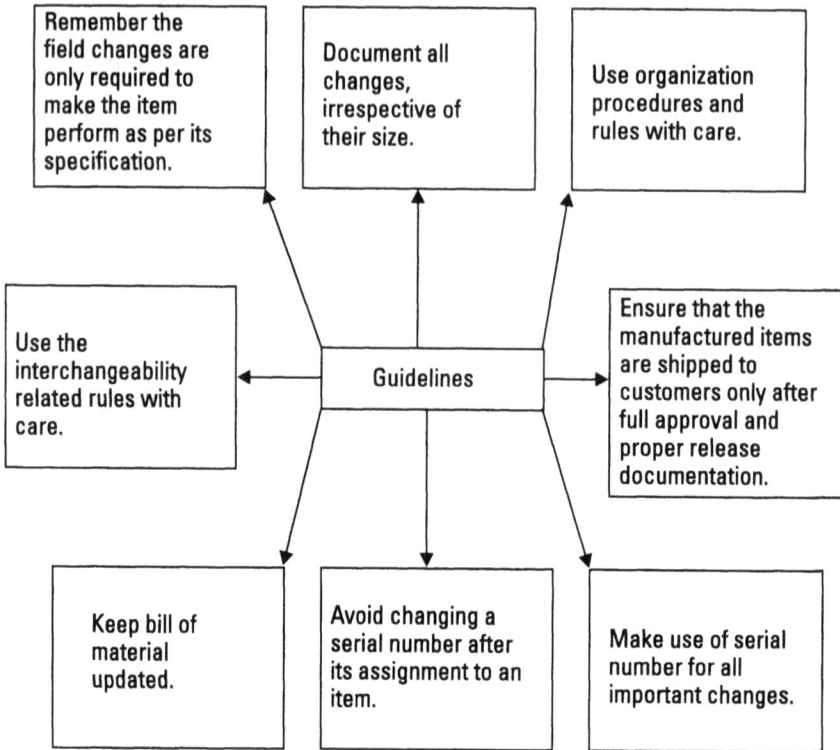

Figure 11.4 Some guidelines for controlling products' configuration and documentation.

- Excessively high cost of changes;
- Inconsistent documentation with the equipment involved.

Past experience indicates that the configuration management approach can be applied effectively to all sizes of organizations or projects, and if implemented effectively its benefits include effective channeling of resources, establishment of objectives for each phase, eradication of redundant effort, and the facilitation of retrieval of accurate data [14].

11.10 Software Configuration Management

Today, complex and sophisticated systems are being developed using software. These systems are made up of a myriad of elements, each of which evolves as it is designed, developed, and maintained. Software configuration management helps to ensure that this evolution is effective and controlled, so that the individual elements fit together in an effective manner to create a coherent whole.

There are probably more publications on software configuration management than on hardware configuration management. However, software configuration management does not differ appreciably from hardware or general configuration management, particularly with respect to objectives, requirements, and functions to be performed.

The fundamental requirements of software configuration management encompass the following four groups [15]:

- Process control;
- Environment management;
- Build management;
- Version control.

Process control includes access control, monitoring, and notification and is a group of policies and enforcement procedures that ensure the software development according to the defined software development methodology. Environment management is the process that chooses and presents the appropriate version of each file to the developer in such a way that permits smooth operation of the development tools.

Build management is a process that concerns the development of software elements, in addition to producing a *bill of materials* that documents each software development's contents. Version control is a process that monitors changes to each file. It augments parallel development by enabling easy merging and branching. Furthermore, each object that evolves in the software development environment ought to be version controlled appropriately.

Some of the key software configuration management planning and organization-related points are listed here [16].

- Ensure that the configuration management plan considers factors such as the project size, the requirements for variants, the types of items to be managed, and the use of third-party software.
- The configuration management requirements may vary from one project to another.

- Manage the configuration management plan in terms of an item and make refinements to it as the project progresses.
- Effectively apply the following two rules in the evaluation of configuration management tools:

  (a) No configuration management tool is a panacea.
  (b) The tool completely satisfies the project needs.

- Aim to keep the configuration management plan structure according to the Institute of Electrical and Electronics Engineers (IEEE) Standard-82890 and ensure that the plan effectively documents every project's approach to configuration management.
- Manage the following two areas of configuration management with care:

  (a) The introduction of most effective configuration management practices and procedures to a wayward project;
  (b) The handing over of software to maintenance people.

Past experience indicates that there are many pitfalls associated with the application of the configuration management approach. Some of the common hazards that must be avoided to achieve effective configuration management results are as follows [17]:

- Too much paperwork;
- Easy or premature acceptance of software developed by others;
- Circumvented and undocumented rules and procedures;
- Availability of only one copy of a configuration-controlled master system;
- Too much time taken to approve the development specification;
- Documentation incompatible with existing programs;
- Software either overcontrolled or undercontrolled;
- No low-level reviews performed prior to the performance of high-level reviews.

11.11 Problems

1. Write an essay on the history of configuration management.

2. Define the following terms:

- Configuration item;
- Configuration control;
- Baseline.

3. What are the reasons for making changes during the design and development phases of a product?

4. List reasons for having a configuration management system.

5. Describe the configuration control board.

6. List at least seven qualities of a configuration manager.

7. What are the key features of successful configuration management?

8. Discuss software configuration management.

9. What are the symptoms of an ineffective application of configuration management?

10. Discuss phases and baselines of a configuration management plan.

References

[1] Monahan, R. E., *Engineering Documentation Control Practices and Procedures*, New York: Marcel Dekker, 1995.

[2] Laine, M. J., and E. C. Spevak, "Configuration Management," *Space/Aeronautics*, November 1966, pp. 74–81.

[3] Feller, M., "Configuration Management," *IEEE Trans. on Engineering Management*, Vol. 16, 1969, pp. 64–66.

[4] Hantz, E. G., and A. E. Lager, "Configuration Management: Its Role in the Aerospace Industry," *Proc. Product Assurance Conference*, 1968, pp. 295–300.

[5] Samaras, T. T., and F. L. Czerwinski, *Fundamentals of Configuration Management*, New York: John Wiley & Sons, 1971.

[6] Dhillon, B. S., *Systems Reliability, Maintainability, and Management*, New York: Petrocelli Books, 1983.

[7] ANSI/EIA-649-1998, "National Consensus Standard for Configuration Management," American National Standards Institute, New York.

[8] DEF STAN 57-50/2, "Configuration Management Policy and Procedures for Defense Material," Ministry of Defense, London, U.K.

[9] BS 6488, "Configuration Management Standard," British Standards Institution (BSI), London, U.K.

[10] ANSI/IEEE-STD-828, "Standard for Software Configuration Management Plans," American National Standards Institute, New York.

[11] Kelly, M., *Configuration Management,* New York: McGraw-Hill, 1996.

[12] Bunker, W. B., "Objectives of Configuration Management," *Defense Industry Bulletin,* September 1967, pp. 1–3.

[13] Seith, W., "Configuration Management in the Navy," *Defense Industry Bulletin,* April 1967, pp. 4–7.

[14] Hajek, V. G., *Management of Engineering Projects,* New York: McGraw-Hill, 1977.

[15] Leblang, D. B., and P. H. Levine, "Software Configuration Management: Which Is It Needed and What Should It Do?" in *Software Configuration Management,* J. Estublier (ed.), Berlin: Springer-Verlag, 1995, pp. 53–60.

[16] Whitgift, D., *Methods and Tools for Software Configuration Management,* New York: John Wiley & Sons, 1991.

[17] Buckle, J. K., *Software Configuration Management,* London: Macmillan, 1982.

12

Total Quality Management

12.1 Introduction

Total quality management (TQM) is a philosophy of pursuing continuous improvement in each process through the integrated efforts of all individuals in the organization.

Although the history of the quality-related efforts may be traced back to ancient times (e.g., Egyptian wall paintings of around 1450 B.C. show evidence of measurement and inspection activity [1]), the roots of the total quality movement only go back to the early 1900s in the time and motion study works of Frederick W. Taylor, the father of scientific management [2–4]. The efforts of such individuals as W. E. Deming, J. Juran, and A. V. Feigenbaum in the late 1940s greatly helped to strengthen the TQM movement [5].

In 1951, the Japanese Union of Scientists and Engineers established a prize named after Deming to be awarded to the organization that demonstrated the most effective implementation of quality policies and procedures [6]. In 1985, an American behavioral scientist, Nancy Warren, coined the term *total quality management* [7]. In 1987, the U.S. government established the Malcolm Baldrige Award for companies demonstrating the most successful implementation of quality-assurance policies and procedures.

In 1988, the first Malcolm Baldrige Award was given to the cellular telephone division of Motorola for its achievement in lowering defects from 1,000 per million to 100 per million during 1985 to 1988 [8, 9]. Also, the division set a goal of reducing defects to four defects per million by 1992. This chapter presents some of the important aspects of TQM.

12.2 TQM-Related Terms and Definitions

This section presents terms and definitions directly or indirectly related to TQM [10–14]:

- *Quality.* This is the totality of characteristics and features of an item or service that bear on its ability to meet specified requirements.
- *Quality program.* This is the documented plans to implement the quality system.
- *TQM.* This is a philosophy, a set of methods, techniques, or tools, and a process whose output results in satisfied customers and continuous improvement.
- *Quality measure.* This is a quantitative measure of the characteristics and features of an item or a service.
- *Quality system.* This is the collective plans, events, and activities that are provided to ensure that an item, service, or process will meet specified requirements.
- *Quality management.* This is the totality of activities involved in determining and achieving quality.
- *Relative quality.* This is the degree of excellence of an item or a service.
- *Quality assurance.* This is all of the planned/systematic actions appropriate to provide satisfactory confidence that an item or a service will meet specified requirements.
- *Quality performance reporting system.* This is the system for collecting and reporting performance statistics concerning product and service.
- *Quality control.* This is the operational methods and the activities that sustain the product or service quality that will meet specified requirements; also the application of such methods and activities.

12.3 TQM and Traditional Quality Assurance Program Versus TQM

The term TQM is composed of three words: total, quality, and management. For its clear understanding, each of these three words is discussed next [15, 16]:

1. *Total.* It means everything and focuses on the need to involve all parties (i.e., inside and outside) of the organization to satisfy

customers. Some of the factors involved in having an effective supplier-customer relationship include customers making suppliers understand their requirements correctly, customers developing their in-house needs systematically, customers regularly monitoring the products and processes of suppliers, and the development of a customer-supplier relationship based on mutual trust and respect.

2. *Quality.* This is deceptively simple, but endlessly complicated in reality, and probably that is why it has numerous definitions. Quality must be viewed from the customer perspective. This view is supported by the result of a survey reported in [15] (i.e., 82% of the respondents categorically stated that quality is defined by the customers and not by the suppliers).

3. *Management.* The approach to management is crucial in determining a company's ability to achieve its set goals and to allocate resources effectively. It means the company must be managed in such a manner that all aspects of business processes (i.e., the *total*) are performed in a way that ensures *quality*. Thus, the foundation of this new type of management system is the involvement of all individuals in the decision making and operation of the company. This clearly indicates that the organizations contemplating the introduction of the TQM approach will have to view their workers with new visions.

All in all, TQM rests on these three items: total, quality, and management, and each of these must have the unwavering attention of the company for TQM to be successful.

In order to show the difference between the traditional quality-assurance program and TQM, both of these areas will be examined with respect to seven factors: objective, quality defined, cost, definitions, customer, decision making, and quality responsibility [6, 9, 16].

Traditional Quality-Assurance Program

- Finds errors;
- Creates manufactured goods that meet specifications;
- Better quality leads to higher cost;
- Product driven;
- Ambiguous understanding of customer or consumer requirements;
- Follows top-down management approach;
- Quality control or inspection department.

TQM

- Prevents errors;
- Manufactures goods suitable for consumer use or application;
- Better quality decreases cost and improves productivity;
- Customer driven;
- Well-defined mechanisms to comprehend and satisfy customer requirements;
- Follows a team approach with groups of workers;
- Involves all individuals in the organization.

12.4   TQM Principles and Components

The concept of TQM is based primarily on the following two principles:

- Customer satisfaction;
- Continuous improvement.

Customers could either be internal or external, and the employment of the "market-in" approach permits a strong customer orientation that recognizes that each and every work process is composed of stages. During each stage, inputs from customers are sought to determine the changes to be made so that the customers' needs are better satisfied [17].

Continuous improvement is a key factor in satisfying the quality challenge.

Past experience shows that excellent companies believe in constant improvement and constant change. Management performs the following two functions with respect to TQM:

- Improving currently used methods and procedures continuously through process control;
- Directing efforts to attain key advances in concerned processes.

Figure 12.1 shows the seven important components of TQM [17]. Each of these components is discussed next.

Management commitment and leadership are critical in the success of TQM, and they are usually achieved only after a thorough understanding of the TQM concept by the senior management. Consequently, the management develops company goals and directions and plays the leadership role in realizing such goals and directions.

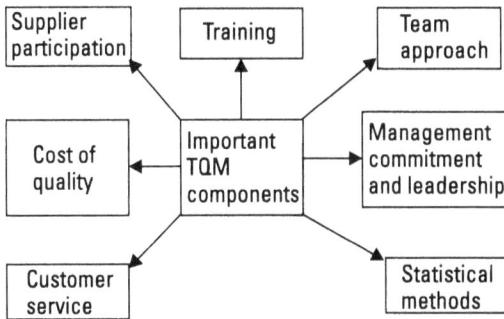

Figure 12.1 Seven important TQM components.

Supplier participation is an important element of TQM because a firm's ability to provide quality goods or services largely depends upon the types of relationships that exist among the parties involved in the process (i.e., the processor, the supplier, and the customer). A key to achieving supplier responsiveness is mutual trust and partnership. All in all, nowadays the increasing trend is that companies are requiring their suppliers to have effective TQM programs as a precondition for potential business [18].

Team approach is another important component of TQM, and its main goal is to get everyone involved with TQM, including vendors, subcontractors, and customers. The size of the quality team may vary from three to 15 members, and membership is voluntary. Nonetheless, team members possess effective knowledge in areas such as brainstorming, cost-benefit analysis, flow charting, statistic-presentation methods, public relations, and planning and controlling projects. The team leader usually has a management background and possesses qualities such as communication skills, presentation skills, group leadership skills, skill in statistical methods and techniques, and skill in group dynamics.

Statistical methods are used for various purposes in TQM applications: to identify and separate quality problem causes; to communicate information in a language that can easily be understood by all team membership; to make decisions on facts based on data instead of opinions of various individuals; and to verify, reproduce, and repeat measurements based on data [17, 19]. Some examples of statistical methods are cause-and-effect diagrams, scatter diagrams, control charts, Pareto diagrams, and graphs and histograms [20–22].

Training is another key component of TQM—a Japanese axiom says that quality begins with training and ends with training [20]. Under TQM, quality is the responsibility of all individuals employed by the company; it means the training effort must be targeted for every hierarchy level of the company. More specifically, it should be tailored accordingly for groups such as engineers, home and field staff, technicians, management, field labor, and support personnel. The training program should cover areas such as team problem solving, rudimentary statistical methods, fundamentals of TQM, interpersonal communication and interaction, cause-and-effect analysis, and quality-measurement cost.

Customer service is another key component of TQM. The application of the TQM approach to the customer area in the form of joint teams normally results in customer satisfaction. The primary objective of forming these teams is to determine customer satisfaction through interactions with customers. These teams establish joint plans, goals, and controls.

The cost of quality is the basic quality measurement tool and in TQM it is used to monitor the effectiveness of the TQM process, justify the cost to doubters, and select quality improvement projects. The cost of quality is expressed by [23]

$$CQ = C_{qm} + C_d$$
$$= C_p + C_a + C_d$$

(12.1)

where

CQ	is the cost of quality.
C_{qm}	is the cost of quality management.
C_d	is the deviation cost.
C_a	is the appraisal cost.
C_p	is the prevention cost.

Two important advantages of the cost of quality are that it is:

- An effective tool to communicate to management the benefits of TQM with respect to dollars;
- An effective tool to raise quality awareness.

12.5 Deming's Approach to TQM

Deming, probably the best-known quality guru, presented the following 14-point approach to TQM [6, 24, 25]:

1. *Establish constancy of purpose for improving services.* The key to this point is the development of a mission statement addressing issues such as long-term corporate objectives, quality philosophy, growth plans, investors, customers, employees, and profit distribution.

2. *Show leadership in promoting change.* This calls for no longer accepting the current levels of mistakes, defects, or delays and alerting all involved individuals to determine why these mistakes, defects, or delays exist and requiring them to work as a team to eradicate such shortcomings.

3. *Stop relying on mass inspection and build quality into the product being manufactured.* More specifically, the mass inspection neither provides any guarantee for quality nor improves it.

4. *Stop the practice of awarding business with respect to price and establish long- term relationships based on actual performance.*

5. *Improve quality, product, and service continuously.* In particular, understand that any process or product variation is bad. There are two kinds of variations: common variations and special variations. The former is beyond the control of the operator, and the latter is within his or her control. The common variations account for approximately 94% of all variations and are under the control of management [26]. The causes of the common variations include inadequate lighting, poor testing of incoming materials, hasty designs, humidity, and noise.

6. *Institute training that embraces modern techniques and approaches.* Carefully consider factors such as identification of goals, identification of objectives, identification of employees, training of supervisors or management, and need analysis in the development of a training program.

7. *Use modern supervisory techniques.* More specifically, this calls for teaching and instituting leadership.

8. *Eradicate fear altogether.* Past experience clearly indicates that people usually find work unpleasant, not because they dislike doing their jobs, but because of the environment under which they have to perform these jobs.

9. *Eradicate barriers between groups/departments altogether and place emphasis on teamwork.*

10. *Eliminate numerical goals, posters, and slogans demanding new levels of productivity without improving the quality of techniques employed.*

11. *Eradicate barriers to worker pride in workmanship.*

12. *Eliminate management by objectives and numerical quotas.*

13. *Promote a dynamic education and self-improvement program.*

14. *Institute appropriate ways and means to accomplish transformation.*

12.6   TQM Methods and Techniques

Over the years professionals working in the area of quality have developed many methods and techniques that can also be used in the application of the TQM concept [16, 24–26]. Reference [26] presents 100 TQM methods and techniques grouped into four areas: management, idea generation, analytical, and data collection, analysis, and display. Tables 12.1 through 12.4 present selective methods and techniques belonging to each of these areas, respectively. This section presents some of these methods and techniques.

Table 12.1

Selective Methods and Techniques Belonging to the Management Area

Method/Technique Name
Kaizen
Pareto diagram
Quality function development
Benchmarking
Quality circles
Zero defects
Affinity diagram
Arrow diagram
Consensus reaching
Objective ranking
Potential-problem analysis
Why-how charting
Contingency planning
Relation diagram
Error proofing (pokayoke)

Table 12.2
Selective Methods and Techniques Belonging to the Idea Generation Area

Method/Technique Name
Brainstorming
Morphological forced connections
Buzz groups
List reduction
Improve internal process plan
Lateral thinking
Mind mapping
Opportunity analysis
Nominal group technique
Multivoting
Imagineering
Snowballing
Idea writing
Suggestion schemes
Rich pictures

12.6.1 Cause-and-Effect Diagram

This diagram was developed by Kaoru Ishikawa in 1943 and is sometimes in the published literature, referred to as the Ishikawa diagram or the Fishbone diagram [27].

As stated earlier, the term *Fishbone diagram* is used because its shape resembles the bones of a fish. More specifically, a cause-and-effect diagram may simply be described as a picture made up of symbols and lines designed to denote a meaningful relationship between an effect and its associated causes. It is used to examine either a "bad" effect or a "good" effect and to take action to rectify the causes and to learn about more causes responsible, respectively.

Pictorially, the right-hand side of the diagram (i.e., the fish head) denotes effect, and the left-hand side denotes all possible causes connected to the central "fish" spine. In developing the Fishbone diagram for the TQM effort, the effect is customer satisfaction and the major causes usually are

Table 12.3
Selective Methods and Techniques Belonging to the Analytical Area

Method/Technique Name
Taguchi methods
Paired comparisons
Cause-and-effect analysis
Force field analysis
Minute analysis
Robust design
Domainal mapping
Solution effect analysis
Failure mode and effect analysis
Fault tree analysis
Evolutionary operation
Stratification

manpower, materials, methods, and machines. In turn, each major cause has numerous subcauses. For example, the subcauses of methods could be knowledge, training, and ability. The following steps are associated with the cause-and-effect diagram:

- Develop statement of the problem.
- Brainstorm to highlight probable causes.
- Establish major cause classifications by stratifying them into natural groupings and process steps.
- Construct the diagram by linking the identified causes under suitable process steps and state the effect or problem in the "fish head box" on the extreme right-hand side of the diagram.
- Make refinements to the cause groups by asking questions such as: What is the reason for the existence of this condition and what causes this?

Some of the advantages of the cause-and-effect diagram are that it is an excellent tool to identify most causes, it presents an orderly arrangement of

Table 12.4
Selective Methods and Techniques Belonging to the Data Collection, Analysis, and Display Area

Method/Technique Name
Hoshin Kanri
Scatter diagrams
Tally charts
Pie charts
Matrix data analysis
Histograms
Bar charts
Check sheets
Cusum charts
Concentration diagrams
Box and whisker plots
Dot plots
Paynter chart
Matrix diagram
Spiderweb diagrams

theories, it is helpful in guiding further inquiry, it is useful in produceing new ideas, and it is useful in identifying possible areas for data collection.

12.6.2 Pareto Diagram

This is a good TQM tool and is named after Italian economist and sociologist Vilfredo Pareto (1848–1923). The Pareto principle, with respect to quality control, simply states that there are always a few kinds of defects in the manufactured products that loom large in frequency of occurrence and severity. The same principle may also be interpreted as that on the average about 80% of the quality costs occur due to 20% of the defects.

Pictorially, the Pareto diagram is a frequency chart in which bars are arranged in descending order from left to right. It provides order to activity.

The Pareto diagram is an extremely powerful tool in TQM work when a team is performing analysis of data because it helps to identify the factors that will have the most impact on the problem and therefore should be addressed first. In short, the Pareto diagram is a useful approach to highlight areas for concerted effort.

12.6.3 Quality Function Deployment (QFD)

QFD was developed by two Japanese professors in 1972, and it may simply be described as a formal process employed for translating customer requirements into technical requirements [24, 28]. QFD ensures the correct deployment of the customer needs throughout the organization (i.e., from product planning stage to field use). More specifically, this method uses a set of matrices to relate consumer needs to counterpart characteristics expressed as technical specifications and process-control requirements. The key QFD planning documents include process plan and quality control charts, a product characteristic deployment matrix, a customer requirements planning matrix, and operating instructions. QFD is also known as *the house of quality.* The following steps are used to build the house of quality:

- Identify customer needs;
- Highlight the process or product characteristics that will meet the customer's needs.
- Relate the customer requirements with counterpart characteristics.
- Evaluate competing products.
- Identify the counterpart characteristics of competing products and develop goals.
- Select counterpart characteristics for use in the remaining process.

The QFD process helps to answer various types of questions: How can the product or process be changed? What does the customer want? Are all wants equal in importance? In what way does an engineering change affect other technical descriptors? In what way does an engineering decision affect customer perception? What is the relationship to process planning, production planning, and parts deployment? Will delivering perceived requirements yield a competitive benefit?

Some of the advantages of QFD are as follows [27]:

- Better customer satisfaction;
- Less engineering design changes;
- Less start-up costs.

Similarly, a principal disadvantage of the QFD is that the exact needs must be identified in complete detail.

12.6.4 Hoshin Kanri (Quality Policy Deployment)

This is a useful method used to delight the customer through the manufacturing and servicing processes by implementing the company quality goals. The method is used when objectives are highlighted at each organizational level through top-down and bottom-up consultation and the global goals of the company have been established as specific targets.

The following factors are associated with the usage of Hoshin Kanri [26, 29]:

- Develop short- and long-term company goals.
- Highlight the goals or objectives that are measurable.
- Identify the crucial processes involved in attaining the goals.
- Require involved groups or teams to agree on performance indicators at appropriate process stages.
- Challenge each process level to require the company to change the quality culture.
- Use organizational goals as measurable objectives to make workers understand the criticality of the quality-improvement process.

The main advantage of this method is that it shows workers the company global goals and where they fit in to achieve these goals.

12.6.5 Check Sheets

The main objective of this useful TQM method is to ensure that the data are collected with care by the operating manpower for process control and problem solving. The collected information is presented in such a way that it can easily be analyzed and used. Usually, the check sheets are designed by the project team, and their designs may vary from one situation to another.

In the design of a check sheet, creativity plays an important role in making check sheets user friendly. Also, it is useful, if possible, to include data in these sheets on location and time. The following steps are involved in developing a check sheet diagram [21, 26]:

- Identify data to be collected.
- Design the check sheet accordingly.

- Test the effectiveness of the check sheet through someone not involved in its design.
- Develop the master check sheet.
- Collect appropriate data.

One important advantage of check sheets is that they present an excellent way to involve people in quality improvement.

12.7 Facts About TQM-Related Organizational Changes and Barriers to TQM Success

Adopting the TQM concept involves various organizational changes within a company. Some of the key facts about organizational changes are as follows [30, 31]:

- Any change will take time.
- Imposed changes will face resistance.
- Changes can be accomplished, but they are difficult.
- Changes may not yield positive results at first.
- Changes may follow course against your intention.
- Changes require total cooperation, participation, and commitment from all management levels.

Over the years, professionals working in the TQM area have identified many barriers to success with TQM. The common ones are shown in Figure 12.2 [30, 32].

12.8 Problems

1. Discuss historical developments in TQM movement.
2. Define the following terms:
   - Total quality management;
   - Quality;
   - Quality program.
3. Compare the traditional quality assurance program with TQM.

Figure 12.2 Important barriers to TQM success.

4. List important elements of TQM.

5. Discuss two basic TQM principles.

6. Describe the Deming approach to TQM.

7. Describe the following TQM methods:

- Quality function deployment;

- Pareto diagram.

8. List at least 14 TQM methods belonging to the management area.

9. What are the important barriers to the success of TQM?

10. Discuss in detail the following two words used in the term TQM:

- Total;

- Management.

References

[1] Dague, D. C., "Quality: Historical Perspective," in *Quality Control in Manufacturing*, Warrendale, PA: Society of Automotive Engineers, 1981.

[2] Rao, A., et al., *Total Quality Management: A Cross Functional Perspective,* New York: John Wiley and Sons, 1996.

[3] Gevirtz, C. D., *Developing New Products with TQM,* New York: McGraw-Hill, 1994.

[4] Goetsch, D. L., and S. Davis, *Implementing Total Quality,* Englewood Cliffs, NJ: Prentice Hall, 1995.

[5] Gevirtz, C. D., *Developing New Products with TQM,* New York: McGraw-Hill, 1994.

[6] Schmidt, W. H., and J. P. Finnigan, *The Race Without a Finish Line: America's Quest for Total Quality,* San Francisco, CA: Jossey-Bass Publishers, 1992.

[7] Walton, M., *Deming Management at Work,* New York: Putnam, 1990.

[8] Van Ham, K., "Setting a Total Quality Management Strategy," in *Global Perspectives on Total Quality,* New York: The Conference Board, New York, 1991.

[9] Madu, C. N., and K. Chu-Hua, "Strategic Total Quality Management (STQM)," in *Management of New Technologies for Global Competitiveness,* C. N. Madu (ed.), Westport, CT: Quorum Books, 1993, pp. 3–25.

[10] ANSI/ASQC A3-1978, "Quality Systems Terminology," Milwaukee, WI: American Society for Quality Control, 1978.

[11] "Glossary and Tables for Statistical Quality Control," Milwaukee, WI: American Society for Quality Control (ASQC), 1973.

[12] Omdahl, T. P. (ed.), *Reliability, Availability, and Maintainability (RAM) Dictionary,* Milwaukee, WI: ASQC Quality Press, 1988.

[13] Hradesky, J. L., *Total Quality Management Handbook,* New York: McGraw-Hill, 1995.

[14] Dhillon, B. S., *Quality Control, Reliability, and Engineering Design,* New York: Marcel Dekker, 1985.

[15] Farquhar, C. R., and C. G. Johnston, *Total Quality Management: A Competitive Imperative,* Report No. 60-90-E, Ottawa, Ontario, Canada: Conference Board of Canada, 1990.

[16] Dhillon, B. S., *Advanced Design Concepts,* Lancaster, PA: Technomic Publishing Company, 1998.

[17] Burati, J. L., M. F. Matthews, and S. N. Kalidindi, "Quality Management Organizations and Techniques," *Journal of Construction Engineering and Management,* Vol. 118, March 1992, pp. 112–128.

[18] Matthews, M. F., and J. L. Burati, "Quality Management Organizations and Techniques," Document 51, Austin, TX: The Construction Industry Institute, 1989.

[19] Perisco, J., "Team Up for Quality Improvement," *Quality Progress,* Vol. 22, No. 1, 1989, pp. 33–37.

[20] Imai, M., *Kaizen: The Key to Japan's Competitive Success,* New York: Random House, 1986.

[21] Kume, H., *Statistical Methods for Quality Improvement*, Tokyo: The Association for Overseas Technology Scholarship, 1985.

[22] Ishikawa, K., *Guide to Quality Control*, Tokyo: Asian Productivity Organization, 1982.

[23] Ledbetter, W. B., "Measuring the Cost of Quality in Design and Construction," Publication 10-2, Austin, TX: The Construction Industry Institute, 1989.

[24] Mears, P., *Quality Improvement Tools and Techniques*, New York: McGraw-Hill, 1995.

[25] Heizer, J., and B. Render, *Production and Operations Management*, Upper Saddle River, NJ: Prentice Hall, 1995.

[26] Kanji, G. K., and M. Asher, *100 Methods for Total Quality Management*, London: Sage Publications Ltd., 1996.

[27] Besterfield, D. H., *Quality Control*, Upper Saddle River, NJ: Prentice Hall, 2001.

[28] Yoji, K. (ed.), *Quality Function Deployment*, Cambridge, MA: Productivity Press, 1990.

[29] Akao, Y., *Hosin Kanri: Policy Deployment for Successful TQM*, Cambridge, MA: Productivity Press, 1991.

[30] Smith, G. M., *Statistical Process Control and Quality Improvement*, Upper Saddle River, NJ: Prentice Hall, 2001.

[31] Hawley, J. K., "Where Is the Q in TQM?" *Quality Progress*, October 1995, pp. 10–13.

[32] Masters, R. J., "Overcoming the Barriers to TQM Success," *Quality Progress*, May 1996, pp. 50–52.

13

Maintenance Management

13.1 Introduction

Ever since the beginning of the Industrial Revolution, maintaining engineering equipment in the field has always been a challenging issue. Today, this problem has become even more pressing because of factors such as the amount of equipment used in the industrial sector, the expected cost effectiveness, and the increase in equipment sophistication, mechanization, and automation. For example, in the U.S. industrial sector alone, the amount of equipment used is so large that over \$300 billion are spent on plant maintenance and operations each year [1].

It means that maintenance management is a very crucial factor in industrial output and even a small improvement in the effectiveness of maintenance effort can result in billions of dollars in savings. For example, a study conducted in the United Kingdom in 1968 indicated that better maintenance practices could have saved approximately £300 million annually of lost production alone due to equipment unavailability [2].

Needless to say, the importance of the maintenance function has increased more than ever before, and good management is the key to its effectiveness. This chapter presents some important aspects directly or indirectly concerned with maintenance management.

13.2 Maintenance Management Terms and Definitions

This section presents terms and definitions directly or indirectly related to maintenance management [3–6].

- *Maintenance.* This is all actions necessary to retain an item in, or restoring it to, a specified state.

- *Capital maintenance.* This is expenditures for the purchase and expansion of plant assets, and it usually includes the installation cost.

- *Maintenance engineering.* This is that activity of equipment maintenance which develops criteria, concepts, and technical needs in the conceptual and acquisition phases to be utilized and maintained in a current status during the operating phase to assure equipment maintenance support.

- *Expense maintenance.* This is expenditures for maintenance and repairs appropriate to the ownership and use of equipment and plant used for manufacturing goods.

- *Maintenance plan.* This is a document that describes the management or technical procedures that will be employed to maintain equipment, and normally it outlines resources, facilities, schedules, and tools.

- *Reliability-centered maintenance.* This is a systematic process employed to find out what must be done to ensure that any physical asset continues to meet its specified functions in its current operating context.

- *Cost of maintenance.* This is cost that includes lost opportunities in uptime, rate, yield, and quality because of nonfunctioning or inadequately functioning equipment or item. It also involves costs associated with equipment or item-related degradation of the environment, property, and personnel safety.

- *Preventable maintenance.* This is work that occurred due to errors of omissions/commission on the part of personnel such as operators, mechanics, plant management, and project engineers.

- *Unplanned downtime.* This is time the equipment is out of action because of unplanned events such as adjustments, setups, breakdowns, and other documented stoppages.

- *Computerized maintenance-management system.* This is a hardware and software system employed to track work orders, equipment histories, and predictive or preventive maintenance schedules.

- *Predictive maintenance.* This is the application of modern measurement and signal-processing techniques to correctly diagnose equipment condition during operation mode.
- *Preventive maintenance.* This is all actions carried out on a planned, periodic, and specific schedule to retain an equipment in a specified operational condition through checking and reconditioning.
- *Corrective maintenance.* This is the unscheduled maintenance or repair actions to put back equipment to a specified condition. It is performed because maintenance personnel or users perceived deficiencies or failures.
- *Maintenance management.* This is the organization of maintenance within an agreed policy.

13.3 Maintenance and Maintenance Engineering Objectives and Good Maintenance Advantages

Although both maintenance and maintenance engineering have the same end goal or objective (i.e., mission-ready system at minimal cost), the environments under which they operate vary significantly. More specifically, maintenance is a function that is often carried out under adverse circumstance and stress, and its primary objective or goal is to quickly restore the equipment to its operational readiness state with the aid of available resources. By contrast, maintenance engineering is an analytical function, and it is methodical and deliberate. Some of the contributing objectives of maintenance engineering are as follows [4]:

- Improve the effectiveness of maintenance operations.
- Improve the effectiveness of the maintenance organization.
- Minimize the degree and frequency of maintenance.
- Minimize the requirements in maintenance skills.
- Reduce the effect of complexity.
- Establish optimum frequency and degree of preventive maintenance to be carried out.
- Improve and ensure maximum use of all concerned maintenance facilities.
- Minimize the volume of maintenance publications.
- Improve the quality of maintenance publications.

- Provide maintenance-related information when required and where required.
- Improve maintenance-related educational programs.

There are many benefits of a good maintenance service, and they may be grouped into five main areas, as shown in Figure 13.1 [7]. These are financial, technical, organizational, human, and customer relations. The financial area includes reduced cost of repairs, extended plant life, reduced production delays, uninterrupted production, improved equipment replacement, less standby plant and spares, and improved quality of production. The technical area includes improved plant condition, improved equipment suitability, buildup of technical data, and improved maintenance schedules.

The organizational area includes coordination between production and maintenance, planning of deliveries, and manpower planning. The human area includes improved safety, improved housekeeping, and less friction or better relations. The customer-relations area includes reliable delivery dates and *showcase* housekeeping.

13.4 Maintenance Department Functions and Organization

A typical maintenance department performs various types of functions. Some of these are as follows [8–10]:

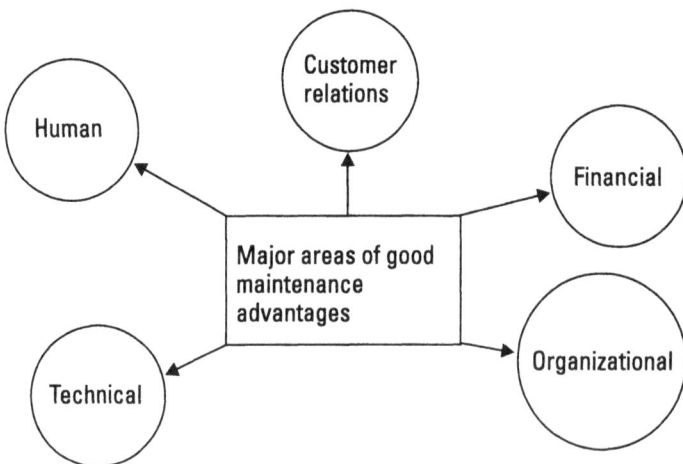

Figure 13.1 Main areas of good maintenance benefits.

- Maintaining records on equipment and services;
- Planning and repairing equipment and facilities to required standards;
- Developing budgets that detail maintenance manpower and material requirements;
- Developing methods for monitoring maintenance manpower activities;
- Performing corrective and preventive maintenance;
- Examining plans for new equipment installation, new facilities, and so on;
- Training maintenance personnel and others to improve their effectiveness in performing tasks;
- Managing inventory to ensure that parts, components, and materials necessary to perform maintenance-related tasks are readily available;
- Developing contract specifications;
- Developing approaches to keep higher management, operations personnel, and others aware of maintenance activities;
- Inspecting work performed by contractors to ensure its compliance with the contract document;
- Developing safety educational programs for maintenance personnel;
- Implementing approaches to enhance workplace safety.

Many factors play an important role in determining the place of maintenance within the plant organization. These include the type of product produced, size, and complexity. In the past the following guidelines have proven to be useful in planning a maintenance organization [9]:

- Fit the organization to the personalities involved;
- Keep vertical lines of authority and responsibility short;
- Establish clear division of authority with minimum overlap;
- Optimize the number of individuals reporting to supervisory personnel.

One of the first considerations in planning a maintenance organization is to decide whether to centralize or decentralize the maintenance function. Past experience indicates that normally centralized maintenance serves quite well small and medium-sized enterprises housed either under one structure or in service buildings located nearby. Some of the advantages of centralized maintenance are that it permits a lesser need for maintenance personnel, a procurement of more modern facilities, a relatively more efficient department than decentralized maintenance, a more effective line supervision, more effective on-the-job training, and a greater utilization of special equipment and facilities and specialized maintenance personnel [11].

Similarly, some of the disadvantages of centralized maintenance are higher transportation costs due to remote maintenance work, more time required getting to and from the work area or job, more challenging supervision due to the remoteness of the maintenance site from the centralized headquarters, and that no one person becomes totally familiar with complex hardware or equipment.

With regard to decentralized maintenance, a maintenance group is assigned to a specific area or some unit. Some of the important factors for having a decentralized maintenance are as follows [6, 11]:

- Closer supervision;

- Less travel time to and from maintenance job;

- The existence of a spirit of cooperation between production and maintenance personnel;

- Better chances for maintenance workers to become familiar with sophisticated equipment or facilities.

The experience gained over the years indicates that in large plants, a combination of both centralized and decentralized maintenance usually works the best. One important reason for this is that the benefits of both of the systems can be achieved with a low number of drawbacks. Thus, no one specific type of maintenance organization is useful for all types of enterprises.

13.5   An Effective Maintenance Management Approach, Maintenance Program Effectiveness Evaluation Questions for Managers, and Maintenance Manager's Qualities and Functions

As making improvements to a maintenance management program is a continuous process, it requires positive attitudes and active involvement from management. An approach for managing a maintenance program effectively is as follows [10]:

- *Identify deficiencies in the system.* This is concerned with identifying existing deficiencies, and this task can easily be accomplished by interviewing maintenance workers and by examining in-house performance indicators.

- *Establish maintenance goals.* This is concerned with setting maintenance objectives or goals by first considering the existing deficiencies and then identifying the targets for improvements.

- *Establish priorities.* This is concerned with prioritizing maintenance projects with respect to savings and merits.

- *Establish performance-measurement parameters.* This is concerned with developing a quantifiable measurement for each established goal (e.g., number of jobs completed per week).

- *Develop short- and long-range plans.* The short-range plan focuses on high-priority goals, normally to be accomplished within a year. The long-range plan is more strategic and highlights important goals to be attained within 3 to 5 years.

- *Document both short- and long-range plans.* This is concerned with documenting both the plans and forwarding copies of the documentation to all involved individuals.

- *Implement plan.* This is concerned with implementing the decided plans with care.

- *Report status.* This is concerned with preparing a report periodically and distributing it to all involved persons. In particular, this report contains, for each short-range goal, data on any actual or potential slippage of the schedule and related causes.

- *Examine progress annually.* This is concerned with reviewing each year's progress with respect to set goals. The following year's short-range plan is developed by considering the long-range plan goals and adjustments to the previous year's planned schedule, costs, and resources.

The U.S. Energy Research and Development Administration surveyed many maintenance managers and then developed a list of 10 questions for maintenance managers to self-evaluate their maintenance effort [12]. These questions were as follows:

1. Are you aware of what facility or equipment and work activity consume the lion's share of the maintenance dollar?

2. Do you know how much time your foreman actually spends at the desk and at the job site?

3. Are you aware of whether your craftspeople use the proper tools and methods to do a job?

4. Are you aware of how the craftspeople spend their time (e.g., travel or delays)?

5. With respect to job costs, can you compare the *should* with the *what*?

6. Do you ensure if maintainability factors are effectively considered in the design of new or modified facilities?

7. Have you balanced your spare parts inventory with respect to carrying cost versus expected downtime losses?

8. Are you aware of whether good safety practices are being followed?

9. Do you have an effective base to measure productivity and if it is improving?

10. Are you providing the craftspeople with the proper quantity and quality of material when and where they need it?

An unqualified "yes" answer to all of these questions means the maintenance-management program is well on the way to meeting the organization's objectives.

In order to perform his or her job effectively, a maintenance manager must possess certain qualities. These include good basic intelligence, communication ability, analytical ability, sound judgment, leadership ability, mechanical aptitude, the ability to make decisions, the ability to delegate, problem-solving ability, planning and organizing ability, the ability to determine what is important, the ability to use time efficiently, and the ability to speak, write, and listen [13]. Some of the important functions of a typical maintenance manager are as follows [14]:

• Supervise the maintenance engineering department.

• Supervise buildings and services maintenance.

• Supervise workshops (e.g., repairs, spares manufacture, and overhauls).

- Plan utility consumption (e.g., electricity, air, waste, steam, and oil).
- Plan and supervise capital work (e.g., installation, construction, and commissioning).
- Direct preventive maintenance activity.
- Direct training and promote safety.

13.6 Effective Maintenance Management Elements

The overall maintenance management function can only be performed effectively if its elements are also executed in an effective manner. It means a careful consideration is necessary of elements such as maintenance policy, job planning, material control, priority and backlog control, data-recording system, work measurement, work order, preventive maintenance, and performance measurement indices [12]. Some of these elements are described next [6, 10, 12].

13.6.1 Maintenance Policy

This is an important element of effective maintenance management and is absolutely essential for providing continuity of operations and a clear understanding of the maintenance-management program. The existence of maintenance policy is essential regardless of the size of a maintenance organization. Normally, maintenance organizations have manuals covering information on topics such as policies, objectives, responsibilities and authorities for all levels of supervision, programs, performance-measuring indexes, useful methods and techniques, and reporting requirements.

If a maintenance organization does not have a maintenance manual, it must at least develop a policy document containing all policy-related information.

13.6.2 Data-Recording System

This plays an instrumental role, directly or indirectly, in the efficiency and effectiveness of the maintenance organization. The system keeps various types of records, including inventory, files, work performed, and maintenance costs.

Inventory records include information such as property number, type, size, procurement cost, manufacturer, date acquired or manufactured, and location of the equipment. Usually, the accounting or stores department provides inventory-related information. File records include such items as drawings, operating and service manuals, and warranties. Maintenance work performed records contain chronological documentation of all preventive maintenance and repairs carried out during an item's service life to date. Maintenance cost records contain historical profiles and accumulations of labor and material costs by item.

Equipment records are used for various purposes, including determining proper levels of preventive maintenance, detecting trends in operating performance, providing information for life cycle costing studies, determining probable causes when troubleshooting breakdowns, helping to justify procurement actions, justifying modification or replacement actions, and performing reliability and maintainability studies at the equipment-design stage.

13.6.3 Work Order

This is used to direct and authorize maintenance persons to perform a required task. A work order should cover all maintenance tasks requested and performed whether repetitive or one time. The work order provides a maintenance manager control of costs and a basis to measure performance for each job. Although the size and type of work order may vary from one organization to another, a work order should collect at least information on areas such as planned start and completion dates, requested completion date, description of and reason for performing the work, item(s) to be affected, required approval signatures, labor and material costs, and work category (e.g., repair, installation, or preventive maintenance).

13.6.4 Job Planning

This is an important element of effective maintenance management and is concerned with formal planning before performing a maintenance job.

Past experience indicates that a number of support-related tasks may have to be performed prior to the commencement of a maintenance job (e.g., procurement of materials and parts; collection and delivery of parts, tools, and materials to the maintenance site; securing of safety work permits; identification of methods and sequencing; and coordination with other departments).

Although the degree of planning required may vary with the type of craft involved and methods used, past experience indicates that generally one planner is required for every 20 craftspersons. Strictly speaking, formal job planning should cover all maintenance jobs. However, usually small true emergency and straightforward jobs are performed in a less formal environment. In most maintenance organizations, 80% to 85% planning coverage is attainable.

13.6.5 Preventive Maintenance

This is an important element of effective maintenance management. There are basically three principal forces that shape the need for and scope of a preventive maintenance program: process reliability, economics, and standards compliance.

The performance of preventive maintenance can help to reduce maintenance cost, provided the effort is not wasted on an item or equipment where the normal expected life without the performance of preventive maintenance exceeds requirements or provided the replacement of noncritical parts is cheaper than the cost of performing periodic inspections and maintenance.

Some of the advantages of performing preventive maintenance are reduction in catastrophic breakdowns, reduction in the manufacturing unit cost, reduction in production downtime, reduction in the need for backup equipment, reduction in maintenance costs, reduction in overtime payments to maintenance personnel, and improved safety for workers [15].

13.6.6 Material Control

In the maintenance activity, the spares and other materials have to be procured, maintained in the inventory, and delivered to the work site. Past experience indicates that on the average, the cost of materials accounts for approximately 30% to 40% of total direct maintenance cost. It means material control is a very important element of effective maintenance management, and the coordination of materials has a key impact on the efficient utilization of maintenance manpower. Ineffective control of materials can lead to problems such as delays, false starts, excessive travel time, and unmet due dates.

One of the most important problems of material control is the decision whether to carry spares in storage or not. In making such decisions, carefully considered factors include procurement lead time, costs of spare parts, criticality of the supported system or item, availability of backup, ability of the vendor or manufacturer to supply spare parts in the future, and reliability of the item or system supported.

Over the years various types of mathematical inventory control models have been proposed to help in decide whether to carry spares in storage. One such model, the inventory control model, is presented here. An idealized diagram of the simplest inventory control model is shown in Figure 13.2.

As per the diagram in Figure 13.2, X number of spare units are ordered at the reorder point and their shipment is received exactly when the old stock is totally depleted. Nonetheless, the model is subject to various other assumptions and [16] provides a comprehensive list of these assumptions.

The following symbols are associated with the model:

- T represents the total cycle time.
- t represents the procurement lead time.
- X represents the total number of spare units acquired at one time.
- Y represents the total number of spare units required annually.
- AC represents the total annual inventory cost.
- PC represents the preparation cost of each order.
- HC represents the annual stockholding cost per spare unit.

Thus, with the aid of Figure 13.2, we write down the following equation for the model:

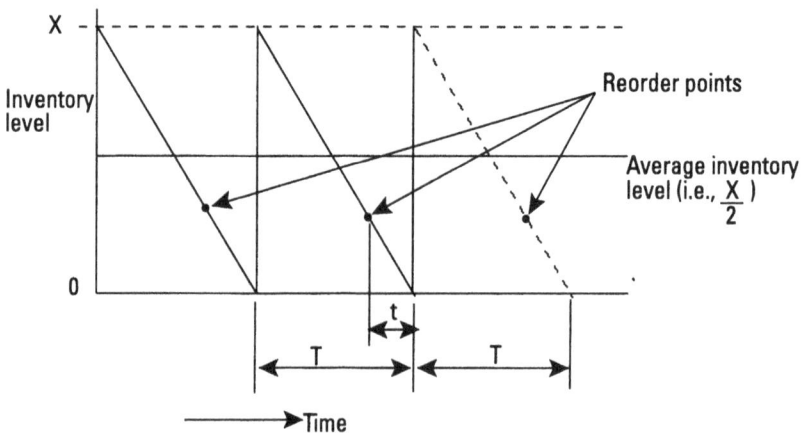

Figure 13.2 Inventory control model idealized diagram.

AC = Annual preperation cost + Annual stock holding cost

$$= \frac{Y}{X}(PC) + \frac{X}{2}(HC) \qquad (13.1)$$

By differentiating (13.1) with respect to X, we get

$$\frac{dAC}{dX} = \frac{HC}{2} - \frac{Y}{X^2}(PC) \qquad (13.2)$$

By setting (13.2) equal to zero and then rearranging, we obtain

$$X^* = \sqrt{\frac{2Y(PC)}{HC}} \qquad (13.3)$$

where X^* is the optimal number of spare units to be ordered at one time.

Using (13.3), we get the following optimum time period, T^*, between orders:

$$T^* = \frac{X^*}{Y} = \frac{1}{\alpha^*} \qquad (13.4)$$

where α^* is the optimal number of annual orders and is given by

$$\alpha^* = \frac{Y}{X^*} \qquad (13.5)$$

Example 13.1

Assume that the maintenance engineering department of a manufacturing company uses 150 spare parts annually for its certain type of equipment. The preparation cost of each order and the annual stock holding cost per spare unit are $60 and $20, respectively.

If the consumption of spare units decreases linearly and the units are replenished instantly (i.e., the moment the stock is depleted), calculate the optimal number of spare units to be ordered at one time.

By substituting the given data into (13.3) yields

$$X^* = \sqrt{\frac{2(150)(60)}{(70)}}$$

$$= 30 \text{ units}$$

It means 30 is the optimal number of spare units to be ordered at one time.

13.7 Performance Measurement Indexes

Various types of indexes are used by management to measure the effectiveness of the maintenance function. These indexes basically show trends with the aid of past data as a reference point. As there is no single index that can effectively reflect the overall performance of maintenance activity, usually a maintenance department uses a number of indexes to measure maintenance effectiveness.

The indexes are quite useful to compare organization, department, and activity performance with other organization, department, and activity performance or industry standards.

The main objective of using these indexes is to encourage maintenance management to improve over the past. Thus, the values of these indexes are plotted periodically to show trends. This section presents a number of indexes taken from the published literature [6, 10, 11, 17–20].

13.7.1 Index I

This index belongs to the family of broad indexes and is expressed by

$$IP_1 = \frac{MC}{IPE} \tag{13.6}$$

where

MC denotes the total maintenance cost.
IPE denotes the total investment in plant and equipment.
IP_1 denotes the index parameter.

In the steel and chemical industries, the approximate average figures for IP_1 are 8.6% and 3.8%, respectively.

13.7.2 Index II

This is another index that also belongs to the family of broad indexes and is defined by

$$IP_2 = \frac{MC}{TS} \qquad (13.7)$$

where

IP_2	denotes the index parameter.
TS	denotes the total sales.

Although the average expenditure for maintenance for all industries is around 5% of the sales, there could be a wide variation among industries. For example, the values for the chemical and steel industries are around 6.8% and 12.8%, respectively.

13.7.3 Index III

This index also belongs to the family of broad indexes and it relates the total maintenance cost to the total output by the organization in question. The index is expressed by

$$IP_3 = \frac{MC}{OP} \qquad (13.8)$$

where

OP	denotes the total output expressed in tons, gallons, or megawatts.
IP_3	denotes the index parameter.

13.7.4 Index IV

This index is known as equipment availability and is expressed by

$$AV = \frac{NH}{TH} \qquad (13.9)$$

where

NH	denotes number of hours each unit of equipment is available to run at capacity.
TH	denotes total number of hours during the reporting period.

AV denotes equipment availability.

Generally, the benchmark value for this index is 96%.

13.7.5 Index V

This index computes maintenance overtime percentage and is expressed by

$$MOP = \frac{MOH}{RMH} \qquad (13.10)$$

where

RMH denotes the total number of regular maintenance hours during the period.

MOH denotes the total number of maintenance overtime hours during the period.

MOP denotes the maintenance overtime percentage.

Usually, the benchmark figure for this index is 5% or less.

13.7.6 Index VI

This is expressed by

$$I_6 = \frac{THWEJ}{THW} \qquad (13.11)$$

where

I_6 denotes the emergency percentage.

$THWEJ$ denotes the total number of hours worked on emergency jobs.

THW denotes the total number of hours worked.

Usually, the benchmark value for this index is 10% or less.

13.7.7 Index VII

This index is expressed by

$$I_7 = \frac{HPPW}{MRMW + HPPW} \tag{13.12}$$

where

MRMW denotes the total man-hours of reactive maintenance work.

HPPW denotes the total man-hours of *preventive maintenance* (PM) or *predictive testing and inspection* (PTI) work.

I_7 is the PM/PTI work percentage.

Usually, the benchmark value for this index is around 70%.

13.7.8 Index VIII

This index is used to control preventive maintenance activity within a maintenance organization. The index is defined by

$$I_8 = \frac{TPPW}{TEM} \tag{13.13}$$

where

TEM denotes the total maintenance time spent for the entire maintenance function.

TPPM denotes the total time spent in performing preventive maintenance.

I_8 is the index parameter.

Past experience indicates that this index should be kept within 20% and 40%.

13.7.9 Index IX

This is a useful index to measure the accuracy of the maintenance budget plan and is defined by

$$I_9 = \frac{AMC}{BMC} \tag{13.14}$$

where

BMC   denotes the total budgeted maintenance cost.

AMC   denotes the actual maintenance cost.

I_9 denotes the index parameter.

13.7.10 Index X

This index is defined by

$$I_{10} = \frac{MLC}{MMC} \qquad (13.15)$$

where

I_{10} denote the index parameter.

MMC   denotes the total maintenance materials cost.

MLC   denotes the total maintenance labor cost.

13.7.11 Index XI

This is expressed by

$$I_{11} = \frac{PJAM}{PJ} \qquad (13.16)$$

where

PJ   denotes the total number of planned jobs.

PJAM   denotes the total number of planned jobs awaiting material.

I_{11} denotes the index parameter.

13.7.12 Index XII

This index is defined by

$$I_{12} = \frac{CPPM}{CPR} \qquad (13.17)$$

where

CPR	denotes the total cost of performing repairs.
CPPM	is the total cost of performing preventive maintenance.
I_{12}	denotes the index parameter.

13.7.13 Index XIII

This is a useful index to relate maintenance cost to manufacturing cost and is expressed by

$$I_{13} = \frac{MC}{TMC} \qquad (13.18)$$

where

| *TMC* | denotes the total manufacturing cost. |
| I_{13} | denotes the index parameter. |

13.8 Problems

1. Define the following terms:
   - Maintenance plan;
   - Maintenance;
   - Maintenance management.
2. List at least 10 contributing objectives of maintenance engineering.
3. What are the main areas of good maintenance benefits?
4. List and discuss at least 10 principal functions of a typical maintenance department.
5. What are the advantages and disadvantages of centralized and decentralized maintenance functions?
6. Write down the five most important elements of effective maintenance management.
7. Write down the important components of a typical work order.
8. What are the important benefits of performing preventive maintenance?
9. Discuss the following two terms:
   - Material control;
   - Job planning.

10. Assume that a maintenance department uses 200 spare parts annually for its certain type of equipment. The preparation cost of each order and the annual stock holding cost per spare unit are $70 and $10, respectively. If the consumption of spare units decreases linearly and the units are replenished instantly (i.e., the moment the stock is depleted), calculate the optimal number of spare units to be ordered at one time.

References

[1] Latino, D. J., *Hidden Treasure: Eliminating Chronic Failures Can Cut Maintenance Costs Up to 60%*, Report, Reliability Center, Inc., Hopewell, VA, 1999.

[2] Kelly, A., *Maintenance Planning and Control*, London: Butterworths and Co. Ltd., 1984.

[3] McKenna, T., and R. Oliverson, *Glossary of Reliability Terms*, Houston, TX: Gulf Publishing Company, 1997.

[4] AMCP706–132, "Engineering Design Handbook: Maintenance Engineering Techniques," Department of Army, Washington, D.C., 1975.

[5] DOD INST. 4151.12, "Policies Governing Maintenance Engineering Within the Department of Defense," Department of Defense, Washington, D.C., June 1968.

[6] Dhillon, B. S., *Engineering Maintenance: A Modern Approach*, Lancaster, PA: Technomic Publishing Company, 2002.

[7] Priel, V. Z., *Systematic Maintenance Organization*, London: Macdonald and Evans Ltd., 1974.

[8] Jordon, J. K., "Maintenance Management," *American Water Works Association*, 1990.

[9] Higgins, L. R., *Maintenance Engineering Handbook*, New York: McGraw-Hill, 1988.

[10] Dhillon, B. S., *Engineering Management*, Lancaster, PA: Technomic Publishing Company, 1987.

[11] Niebel, B. W., *Engineering Maintenance Management*, New York: Marcel Dekker, 1994.

[12] *Maintenance Managers Guide*, Report No. ERHQ-0004, Energy Research and Development Administration, Washington, D.C., 1976.

[13] Eschner, L. F., "Developing Maintenance Managers," in *Modern Maintenance Management*, E. J. Miller and J. W. Blood (eds.), New York: American Management Association, 1963, pp. 73–83.

[14] White, E. N., *Maintenance Planning, Control, and Documentation*, London: Gower Press Ltd., 1979.

[15] Higgins, L. R., and L. C. Morrow (eds.), *Maintenance Engineering Handbook*, New York: McGraw-Hill, 1983.

[16] Riggs, J. L., *Production Systems: Planning, Analysis and Control*, New York: John Wiley and Sons, 1981.

[17] Stoneham, D., *The Maintenance Management and Technology Handbook*, Oxford: Elsevier Science Ltd., 1998.

[18] Hartmann, E., et al., *How to Manage Maintenance*, New York: American Management Association, 1994.

[19] Westerkamp, T. A., *Maintenance Manager's Standard Manual*, Paramus, NJ: Prentice Hall, 1997.

[20] "Reliability Centered Maintenance Guide for Facilities and Collateral Equipment," National Aeronautics and Space Administration (NASA), Washington, D.C., 1996.

14

Warranties, Ethics, and Legal Factors

14.1 Introduction

Today warranties have become an essential element of product marketing in all three sectors (i.e., consumer, commercial, and government). Furthermore, they are driven by law in all these three sectors [1]. A survey of 369 U.S. manufacturers revealed that only one percent of them had no documented or written warranties on the items they manufacture [2]. Furthermore, the survey also revealed that the average cost of a warranty claim was less than 2% of the sale price of an item manufactured.

The history of ethics may be traced back to the ancient times; for example, Aristotle in the fourth century B.C. wrote the Nicomachean Ethics, in which he sought to ascertain answers to the following questions: what is the function or aim of man? what is the good of man? and what will bring man happiness? [3].

In modern context, ethics may simply be described as an area of study involving what is acceptable and what is not acceptable in association with professional and personal obligation and moral duty [4]. Over the past three decades, U.S. industry has witnessed an increasing growth of interest in the development and teaching of corporate ethics. For example, in 1979 a study involving Fortune 500 companies and the top 150 service firms revealed that 73% had some sort of written standards of ethics and 50% of these were 5 years old or less [5]. Furthermore, another study involving 2,000 companies, conducted a decade later, reported that 90% of the companies had standards of ethical conduct.

Engineers and their associates have to deal with various types of legal factors in the practice of their profession. The legal factors involved in engineering could be formal or informal requirements, and the protection of property through the use of copyright, patents, design registration, or trademark is well documented with the associated legal aspects clearly defined.

The branch of law governing the liability of an item's manufacturer is referred to as product liability and the past three decades have witnessed an alarming increase in product liability suits. For example, in 1976, it was estimated that consumer-initiated lawsuits in the United States were around 50,000, while the prediction for 1980 was 1 million [6, 7].

This chapter presents various important aspects of warranties, ethics, and legal factors.

14.2   Reasons for the Warranty Needs and Basic Options to Meet Warranty Obligations

There are many reasons for having a warranty, and some of these are as follows [8]:

- To gain prompt acceptance of the manufactured item;
- To force manufacturers to provide effective product documentation and services;
- To make manufacturers responsible for eradicating faults in their products;
- To enhance the marketability of manufactured items;
- To encourage manufacturers to produce better products;
- To assure buyers that the manufactured products will meet at least their contractual specifications;
- To encourage better monitoring of product performance;
- To allow customers to gain proficiency in operating the product during the no-charge warranty period;
- To expedite payments from customers for sold products;
- To assure that the technical services will be readily available when required.

There are various ways in which product manufacturers can meet their warranty obligations. For example, basic options to meet warranty obligations for electrical and mechanical products can be as follows [2]:

- Replace the faulty parts in items or compensate buyers so that they can purchase such parts.
- Compensate for the replacement of faulty parts and installation or repair cost of such parts.
- Compensate for the service labor to repair defective parts or to replace defective parts with new ones.
- Replace the defective product altogether.

14.3 Government, Commercial, and Consumer Warranties

Government warranties cover a wide range of applications in the United States (e.g., Department of Defense Systems, Department of Transportation Systems, and Department of Energy Systems).

Commercial warranties cover any goods sold between merchants and need not be the same for all buyers. Subsequent to any negotiations, the individually signed agreements or contracts become integral parts of the purchase agreement. In the United States, commercial warranties are subject to the *uniform commercial code* enacted in each state. A typical example of goods sold between merchants is jet airplanes manufactured by aircraft companies and procured by airlines [1].

Consumer warranties cover consumer products and are the same for all buyers. A consumer product may simply be described as any tangible personal property distributed in commerce and is generally utilized for personal, family, or household purposes. In the United States, consumer warranties are subject to the uniform commercial code enacted in each state and the Magnuson-Moss Federal Trade Commission Improvement Act.

14.3.1 Consumer and Government Warranties' Comparison

In the consumer economy sector, the sale price is increased to cover the manufacturer's cost of replacing or repairing the item during the warranted period at the time of purchase. Furthermore, an additional warranty contract may also be purchased to cover repairs or replacement after the expiration of the original warranty period. By contrast, in the government sector, as the cost of warranty cannot be included into the item selling price, it is priced and negotiated separately. However, there are many similarities with respect to requirements between consumer and military warranties. For example, both warranties define factors such as what is to be covered, the extent of warranty coverage, acceptable use conditions and environments, supplier's remedies under the warranty, and the length of time the warranty is in effect.

Some of the comparable factors concerning commercial and military warranties are as follows [1]:

Commercial

- Market research is quite extensive.
- Requirements are self-determined.
- Service is factory authorized.
- User environments are orderly.
- Product is manufactured prior to sale.

Military

- Market research is limited or does not occur.
- Requirements are customer specified.
- Services are performed by users.
- User environments are hectic.
- Product is manufactured after sale.

14.3.2 Consumer and Commercial Warranties' Comparison

In the United States, full consumer warranties under the Magnuson-Moss Federal Trade Commission Improvement Act place various obligations on the product manufacturer including, but not limited to, providing remedies for claims within a reasonable time frame at no cost or charge and refunding the procurement cost minus a reasonable amount of depreciation if it is impossible to rectify the defect within a reasonable number of attempts [1]. By contrast, commercial warranties subject to the uniform commercial code are usually not as restrictive to the manufacturer as those subject to the federal Magnuson-Moss Act. However, if more than 10% of the sales of commercial items or products falls in the consumer area, the items or products become categorized as consumer items or products and thus are subjected to the Magnuson-Moss Act.

14.4 Warranty Components and Management

A written warranty contains various components. For example, the important components of consumer and commercial warranties are as follows [1]:

- The coverage period;
- Manufacturer's responsibilities with respect to defects or obligations;
- Remedies to be executed within a stated time in the event of a failure;
- Warrantor's name and address;
- The items covered by the warranty;
- Countries in which the warranty applies;
- Purchaser's responsibilities (e.g., basic maintenance and the bearing of costs);
- Identification of party to whom the warranty is directed (e.g., original buyer only or any owner within the specified time period);
- Exceptions and exclusions from the terms of warranty;
- The time period for which the warrantor or manufacturer will carry out obligations specified in the warranty;
- The statement of warrantor's action in the event of a problem—conforming with the written warranty at whose expense and for what periods of time;
- The time within which, after the reporting of a problem (i.e., defect, failure, or malfunction), the warrantor will carry out obligations specified in the warranty document;
- Applicable disclaimers;
- Steps for the purchaser to follow to obtain performance of any remedy as per the warranty, including a clear identification of authorized individuals to carry out the specified obligation.

Warranty management plays a key role in the success of warranties provided by the manufacturers on their products. Past experience indicates that the majority of companies assign the warranty administration responsibility either to product or technical service function, and only a small percentage of companies have a distinct division to handle warranties. However, various factors may dictate the positioning of warranty administration responsibility within a company structure: the type of product manufactured, the frequency and type of claims, the company's organizational structure, the importance given by customers to warranty-protection issues, and the characteristics of a company's distribution system [2].

Under certain conditions, the warranty administration responsibility is assigned to the following units [2]:

- *Separate department.* This happens when the cost or number of claims is fairly high.

- *Customer service.* This happens when the claims are relatively simple and straightforward and their number is small.

- *Product or technical service.* This happens when the claims require a certain degree of technical knowledge for processing.

14.5 Reasons for the Growth in Warranty Claims, Predictions for Warranty Problems in the Future, and Warranty Cost Estimation Models

There could be many reasons for the growth in warranty claims. Some of the common ones are as follows [7]:

- Deterioration in the manufactured items' quality;

- Better awareness among customers concerning the warranty obligations of manufacturers;

- Bigger product volume (i.e., the greater the volume, the higher the chances for more warranty claims);

- More complex, sophisticated, and service-sensitive manufactured items.

- Product misuse by the user or customer;

- Unsatisfactory maintenance in the user facilities due to inadequately trained and qualified maintenance manpower;

- Inflationary pressures.

Over the years various factors have been put forward concerning future warranty problems. Some of these factors are as follows:

- Increase in user or customer expectations;

- Increase in product liability;

- Increase in litigation activity;
- Government rules and regulations;
- Increase in cost in areas such as labor and material;
- Short supply of qualified product-service personnel;
- The training of users and dealers with respect to proper usage of manufactured items;
- Warranty starting date record keeping.

14.5.1 Warranty Cost-Estimation Models

Over the years many warranty cost estimation models have been developed [9–12]. This section presents two such models.

Model I

This mathematical model uses the following equation to estimate contractor or manufacturer warranty cost [11]:

$$CWC = FWC + \lambda (TOH)(ARC) \qquad (14.1)$$

where

CWC is the contractor or manufacturer's warranty cost.

FWC is the contractor or manufacturer's fixed warranty cost.

λ is the hourly failure rate of the warranted item during the warranty period.

ARC is the contractor or manufacturer's average repair cost associated with the warranted item.

TOH is the total operating hours of the warranted item during the warranty period.

Model II

This model uses the following equation to estimate warranty administration cost [12]:

$$WAC = \left(C_{ap}\right)\left(MM_{pm}\right)\left(T_{wa}\right) \qquad (14.2)$$

where

WAC	is the total warranty administration cost for item under consideration.
T_{wa}	is the length of warranty administration period expressed in months.
C_{ap}	is the monthly average cost of an administration person.
MM_{pm}	is the total number of administrative man-months per month.

14.6 Need for Ethics, Ethical Concerns, and an Example of Unethical Behavior

At the start of the twentieth century there were no airplanes, television sets, radios, spacecraft, telephones, atom bombs, or automobiles. In fact, most people traveled within a radius of a few miles of their homes, and the contemplation of man ever traveling to the Moon was a far-fetched fantasy. The technological revolution of the twentieth century has created a social revolution, and in turn an understanding of the relationship between these two is critical to the subject of ethics in technology and engineering [5]. Consequently, the need of ethics in engineering and technology has become an important issue.

There are a wide range of areas in which ethical problems may arise for engineers, including advertising products and services, kickbacks, trade secrets, bidding for contracts, patents, and cartels. Some of the ethics-related questions concerning areas such as these are as follows [4]:

- Are the kickbacks ethical?
- Is it ethical to take advantage of company's facilities for personal use?
- Is it ethical to use other companies' trade secrets to avoid bankruptcy of your employer?
- Is it ethical to promote yourself as available to perform various types of engineering jobs?
- Is it ethical to fix the price of items manufactured by your organization with competitors?
- Is it ethical to "borrow" a patented design for use after patent holder's death?
- Is it ethical to design an item if you are not a licensed engineer?

- Is it ethical to design an item that can directly or indirectly harm humans?

- Is it ethical to make changes to the test data on items at the insistence of your employer?

- Is it ethical to inform others when another colleague is doing something that in your opinion is bad engineering?

Over the years, there have been many well-publicized examples of unethical behavior of engineers. One such example is the resignation of Spiro T. Agnew, vice-president of the United States, on October 10, 1973, because of bribery and tax evasion charges relating to his previous position as county executive of Baltimore County, Maryland [5]. Agnew was a civil engineer and lawyer, and during the period 1962–1966 he awarded many public works contracts to architectural or engineering firms from which he received kickbacks [13].

14.7 General Guidelines for Ethical Behavior and Ethics Code for Engineers

There are many general guidelines for ethical behavior available in published literature. Some of these are as follows [3]:

- Obey the laws of the land in an effective manner.

- Follow your conscience effectively.

- Follow logic and reason with care.

- Strive consciously for happiness.

- Follow the accepted customs and ideas with care.

- Follow the example of great persons consciously.

- Follow the Bible and religious authority, but with care.

All of the major professional societies in the United States have adopted their own codes of ethics to guide and support their members. Consequently, there are many codes of ethics within the engineering profession itself (e.g., the code of ethics of the Institute of Electrical and Electronics Engineers, the American Society of Mechanical Engineers, and the American Society of Civil Engineers).

The code of ethics for engineers developed by the National Society of Professional Engineers may be divided into three distinct groups as follows [5, 13]:

- Basic canons;
- Rules of practice;
- Obligations.

Each of the above categories is described next.

14.7.1 Basic Canons

In the fulfillment of professional duties, engineers should do the following:

- Carry out services in areas of their specialties only.
- Do not solicit professional employment in an improper manner.
- Give paramount attention to public health, safety, and welfare in the performance of their assigned professional duties.
- Act in professional issues for employers and others as faithful trustees or agents.
- Make public statements in a truthful and objective manner.

14.7.2 Rules of Practice

There are five rules as follows:

- Engineers shall carry out services in areas of their specialties only. This includes not signing documents in areas in which they lack competence; undertaking assignments in areas of their specialties only; and while they may accept an assignment outside their fields of competence, they must restrict their services to those areas of the project in which they possess competence.
- Engineers shall not solicit professional employment in an improper manner. This includes not falsifying their academic or professional qualifications and not offering any gift or other valuable consideration to secure work.
- Engineers shall give paramount attention to public health, safety, and welfare in the performance of their assigned professional duties.

This includes approving only those engineering documents that are safe for public health, property, and welfare; recognizing their primary obligation to protect the safety, health, property, and welfare of the public; and not revealing facts, data, or information obtained in a professional capacity without the permission of employers or others unless required by law.

- Engineers shall act in professional issues for employers and others as faithful trustees or agents. This includes disclosing all known or potential conflicts of interest to their employers or clients, not soliciting or accepting financial or other consideration from contractors or their agents, and not soliciting or accepting a professional contract from a government body in which a senior official of their company serves as a member.

- Engineers shall make public statements in a truthful and objective manner. This includes being truthful and objective in professional documents, statements, or testimony, and issuing no statements, criticisms, or arguments on technical matters that are either paid for or inspired by interest groups or parties.

14.7.3 Obligations

There are a total of 11 professional obligations of engineers. They are as follows:

1. Engineers shall strive always to serve the public interests.

2. Engineers shall accept personal responsibility without any hesitation whatsoever for all professional activities performed.

3. Engineers shall not be influenced by conflicting interests when performing day-to-day professional duties.

4. Engineers shall uphold the principle of satisfactory payment for those involved in engineering activity.

5. Engineers shall be guided in their day-to-day professional dealings by the highest possible standards of integrity.

6. Engineers shall not undertake any practice that may deceive the public or discredit the engineering profession.

7. Engineers shall always give credit for engineering effort to deserving individuals and will effectively recognize the proprietary interests of other people.

8. Engineers shall not pass on confidential information relating to business affairs or technical processes of any current or former employer or client without his or her permission.

9. Engineers shall work together in extending professional effectiveness through interchanging relevant experience and information with other engineering professionals and students, and engineers will provide adequate opportunity for the professional development of engineering professionals under their wings.

10. Engineers shall not injure, through any means whatsoever, the professional reputation, employment, practice, or prospects of fellow engineers, nor in an indiscriminate manner criticize fellow engineers' efforts.

11. Engineers shall not compete in an unfair manner with fellow engineers by attempting to obtain employment or advancement of professional engagements through factors such as criticisms and other questionable means.

14.8 Limitations of Ethics Codes

Most of the ethics codes are limited in several ways, and these limitations restrict the codes so that they serve as general guidance only. It means engineers must exercise a personal moral responsibility rather than expect these codes to find solutions to their moral problems. The four major limitations associated with ethics codes are shown in Figure 14.1 [13].

General and vague wording means that because the ethics codes are restricted to general and vague wording, they cannot simply and straightforwardly be applied to all situations. *Different entries* means that it is quite easy

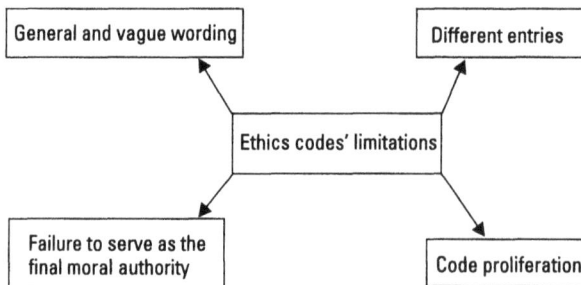

Figure 14.1 Limitations of major ethics codes.

for different entries in ethics codes to conflict with each other. Furthermore, usually these codes do not provide any guidance in terms of entry priority, thus they create moral dilemmas.

Failure to serve as the final moral authority means that these codes cannot serve as the final moral authority for professional conduct [14]. Furthermore, to accept the code of ethics of a professional society as the final moral word would lead to some kind of ethical conventionalism [13]. *Code proliferation* refers to the existence of a large number of codes, thus less uniformity among them leads to poor effectiveness. In the past, some professional societies have tried to develop a unified code of ethics but, unfortunately, thus far to no avail.

14.9 Product Liability and Patents

Today, product liability is an important issue to engineering professionals, as its cost is increasing at an alarming rate. Approximately 25% of product litigation accounts for engineering negligence [15]. In engineering design, several factors could be the basis for negligence: concealment of design dangers, failure of the design to satisfy applicable standards, failure to provide appropriate safety devices as part of the product [16]. Also, it is estimated that approximately 60% of the product liability cases in the United States involve the failure to provide suitable danger warning labels.

Under the strict liability theory, in the United States, the plaintiff needs only to prove the following three factors:

1. The defect in the product or item was unreasonably dangerous.
2. The product defect was responsible for the injury.
3. The defect was present at the time the product or item left the defendant's facility.

In the United States, a product manufacturer may be liable on the basis of the following factors [6, 7]:

- Poor quality control and testing caused manufacturing defects;
- Poor manufacturing, distribution, and sales record keeping;
- Poor labeling concerning possible danger and proper usage;
- Poor failure and user complaint data collection;
- Selling the packaged product in incomplete and dangerous form;

- Susceptibility of the packaged product to safety-related handling damage;
- Susceptibility of instructions to detachment from the packaged product prior to sale.

In order to counteract negligence charges, product manufacturers have developed many points and these points fall into five basic areas: nondetectability of the fault, consumer negligence, state of the art, statute of limitations, and product compliance with necessary standards and regulations. The following guidelines, if carefully followed during design and manufacturing, can help to minimize potential product-liability problems [15]:

- Follow closely government and industry standards.
- Test products or items effectively before their release for sale.
- Document with care activities such as design, manufacturing, testing, and quality control.
- Use as much as possible the improved quality-control methods with the goal of lowering product-liability problems.
- Carefully design the product warning labels and the user's manual as an integral part of the design process. Furthermore, use international warning symbols as much as possible and involve people from manufacturing, engineering, marketing, and legal departments when designing symbols and labels.

Probably, the largest pool of technical information in the world is the U.S. Patent System with over 4 million patents. In the published sources, only about 20% of the technology contained by these patents is available. The U.S. Patent System began in 1790, and a patent may be awarded for a maximum of 17 years under various classifications: machines, processes, human-made micro-organisms, matter compositions, and manufactured articles [15, 16].

There are three common criteria for awarding a patent:

- The usefulness of the invention;
- The invention ingenuity;
- The newness or novelty of the invention.

There are several points concerning patent ownership: a patent is assignable to another individual, a patent is transferable to a beneficiary in a will, and a patent prevents other people from using, selling, or manufacturing the covered item without obtaining express permission [17, 18].

14.10   Copyrights and Trademarks

Copyright may simply be described as the sole, absolute legal right to publish a tangible expression of work of literary or artistic nature, and it protects against the unauthorized copying of such work by other people. The period covered by a copyright in the United States begins from the time of the work's creation to 50 years after the death of the creator. The copyright law of the United States covers four types of original work: literary, graphic, pictorial, and sculptural [19]. More specifically, with respect to engineering design, the law covers first-time written engineering specifications, drawings, models, and sketches. All in all, under the U.S. copyright law, an idea cannot be copyrighted but its tangible expression can be.

A trademark may simply be described as a symbol, letter, or word that identifies certain items or their origin. In the course of trade, the trademarks are useful to make a distinction between similar items. The other advantages of trademarks are that they can create an image for a particular good and they can create goodwill and consumer preference. Some of the qualities of a good trademark are as follows [16]:

- Very distinctive;

- Easy to remember;

- Easy to pronounce;

- Attractive in sound and appearance;

- Easy to reproduce;

- Simple and short;

- Easy to spell and write;

- Suggestive of good qualities;

- Does not mislead;

- Free of offensiveness to the eye and ear.

Under the U.S. Trademark Act of 1946, an application must be submitted to the Patent and Trademark Office to register a trademark. The successful registration lasts for a period of 20 years and it may be renewed before expiration.

14.11   Problems

1. Write an essay on ethics.
2. What are the reasons for the warranty needs?
3. Describe the following types of warranties:
   - Commercial;
   - Government;
   - Consumer.
4. Compare commercial and military warranties.
5. List the important elements of consumer and commercial warranties.
6. What are the common reasons for the growth in warranty claims?
7. Give an example of unethical behavior by an important engineer.
8. List at least seven general guidelines for ethical behavior.
9. What are the limitations of ethics codes?
10. List at least seven factors for which a product manufacturer in the United States could be liable.
11. Discuss the following terms:
    - Patent;
    - Copyright;
    - Trademark.

References

[1]   Brennan, J. R., *Warranties: Planning, Analysis, and Implementation,* New York: McGraw-Hill, 1994.

[2]   McGuine, P. E., *Industrial Product Warranties: Policies and Practices,* New York: The Conference Board, Inc., 1980.

[3]   Mantell, M. I., *Ethics and Professionalism in Engineering,* New York: Macmillan, 1964.

[4]   Walton, J. W., *Engineering Design,* New York: West Publishing Company, 1991.

[5] Ertas, A., and J. C. Jones, *The Engineering Design Process*, New York: John Wiley and Sons, 1993.

[6] Kolb, J., and S. S. Ross, *Product Safety and Liability*, New York: McGraw-Hill, 1980.

[7] Dhillon, B. S., *Engineering Management*, Lancaster, PA: Technomic Publishing Company, 1987.

[8] Flottman, W. W., and M. R. Worstell, "Mutual Development, Application and Control of Suppliers Warranties, American Airlines and Litton Systems-Aero Product Division," *Proc. Annual Reliability and Maintainability Symposium*, 1977, pp. 213–221.

[9] Gates, R. K., R. S. Bicknell, and J. E. Bortz, "Quantitative Models Used in the RIW Decision Process," *Proc. Annual Reliability and Maintainability Symposium*, 1977, pp. 229–236.

[10] Balaban, H., and R. Retterer, *Guidelines for Application of Warranties to Air Force Electronic Systems*, ARINC Research Corporation, RADC Report TR-76-32, Rome Air Development Center (RADC), Griffiths Air Force Base, Rome, March 1978.

[11] Balaban, H. S., and M. A. Meth, "Contractor Risk Associated with Reliability Improvement Warranty," *Proc. Annual Reliability and Maintainability Symposium*, 1978, pp. 123–129.

[12] Isaacson, D. N., S. Reid, and J. R. Brennan, "Warranty Cost-Risk Analysis," *Proc. Annual Reliability and Maintainability Symposium*, 1991, pp. 332–339.

[13] Martin, M. W., and R. Schinzinger, *Ethics in Engineering*, New York: McGraw-Hill, 1989.

[14] Ladd, J., "The Quest for a Code of Professional Ethics," in *AAAS Professional Ethics Project: Professional Ethics Activities in the Scientific and Engineering Societies*, R. Chalk, M. S. Frankel, and S. B. Chafer (eds.), Washington, D.C.: American Association for the Advancement of Science, 1980, pp. 154–159.

[15] Dieter, G., *Engineering Design*, New York: McGraw-Hill, 1983.

[16] Ray, M. S., *Elements of Engineering Design: An Integrated Approach*, Englewood Cliffs, NJ: Prentice Hall, 1985.

[17] Pressman, D., *Patent It Yourself? How to Protect, Patent, and Market Your Inventions*, New York: McGraw-Hill, 1979.

[18] Burge, D. A., *Patent and Trademark Tactics and Practices*, New York: John Wiley and Sons, 1979.

[19] Dhillon, B. S., *Engineering Design: A Modern Approach*, Chicago, IL: Irwin, 1996.

15

Reengineering

15.1 Introduction

In today's dynamic industrial sector, reengineering has become an important issue. For example, a survey of large corporations indicates that approximately 90% of the companies were involved in business process–reengineering projects and many others were seriously planning to execute such projects [1, 2]. Usually the organizations involved in reengineering target their efforts on areas such as development, manufacturing, logistics, distribution, and customer support [3].

In the mid-1950s the Japanese company Toyota revolutionized production processes and delivery capabilities, thus paving the path to reengineering. During the period of the 1973 oil embargo, other Japanese companies were quick to learn the process-oriented concepts of Toyota and started to convert to process-driven production. By 1983, the executive suites of the largest companies in the West were well aware of the basic principles of the Toyota production system known as the *just-in-time* (JIT) manufacturing.

Although some organizations claimed to have performed business process engineering well before 1990, the term *reengineering*, coined by M. Hammer and J. Champy, became popular in 1993 [4–6]. This chapter presents important aspects of reengineering.

15.2   Reengineering Terms and Definitions

Some of the terms and definitions[1] directly or indirectly concerned with reengineering are as follows [4–8]:

- *Reengineering.* This is the examination and alteration of an item or a system under consideration to reconstitute it in some new form and for the subsequent implementation of that form.

- *Software reengineering.* This is the examination and alteration of an existing item or system to reconstitute it in a new form and encompasses a combination of subprocesses such as redocumentation, reverse engineering, restructuring, retargeting, and forward engineering.

- *Redocumentation.* This is the process of analyzing the system or item to develop and produce support documentation in forms such as user manuals and reformatting the source code listing of the systems.

- *Restructuring.* This is the process of transforming the item or system from one form to another at the similar relative abstraction level and at the same time keeping the subject system's external functional behavior intact.

- *Forward engineering.* This is a set of engineering activities that makes use of products or artifacts derived from legacy software and new needs to manufacture or produce a brand-new target system or item.

- *Business process reengineering.* This is the fundamental rethinking and radical redesign of business processes to achieve tangible or dramatic improvements in vital and contemporary measures of performance, such as service, speed, quality, and cost.

- *Data reengineering.* This refers to the tools that conduct all of the associated reengineering functions with source code (forward engineering, redocumentation, translation, retargeting, restructuring, and reverse engineering) but act specifically upon data files.

15.3   Reengineering Facts and Figures

This section presents facts and figures directly or indirectly concerned with reengineering:

---

1.   It is to be noted that some of these definitions may convey meanings from the software perspective or they may be applicable to the software reengineering area only.

- Florida Power and Light Company achieved a reduction in power outage per customer to 32 minutes, compared to 7 hours by its competitor, through reengineering [9].

- CIGNA, a leading provider of insurance and related financial services in the United States, reported that each $1 invested in reengineering generated $2–$3 in returned benefits [10].

- Corning Asahi Video (CAV) Products won *Computerworld's* Annual Reengineering Team of the Year Award for completing a 15-month project costing $570 million that resulted in halving fulfillment time and reducing per order ordering costs for CAV by 75% [11].

- Digital Equipment Corporation successfully eliminated 450 positions through a reengineering project by consolidating 55 accounting groups into just five [9].

- Progressive Insurance, through reengineering, reduced time spent in settling claims from 31 days to just 4 hours [9].

- During the period 1987–1992, Banca di America e di Italia (BAI) doubled its revenue and attributed 24% of the increase to its reengineering effort [12].

- Pacific Bell's first reengineering project was known as "Centrex Provisioning," in which the company reported 36%–50% reduction in cost and over 20% reduction in errors [13].

- C. R. England and Sons, through reengineering, was able to reduce its cost of sending an invoice a to mere $0.15 compared to the average cost of $5.10 that was incurred during the period 1989–1991 [9].

- AT&T Capital Leasing Services, through reengineering, was able to increase its sales by 20% and decrease its credit approval time by 39% [14].

- Rank Xerox (U.K.), through reengineering, reduced its order delivery time from 33 days to 6 days [9].

- Inter-Mountain Health Care of Salt Lake City, through reengineering, reduced infection rates by half, resulting in an annual saving of approximately $750,000, and cut its adverse drug reaction cost by $900,000 per year [15].

15.4   Reasons for Reengineering and When to Reengineer

There could be various reasons for reengineering, including the ones listed next [9, 16]:

- To increase profit;
- To introduce zero technology;
- To achieve the appropriate level of integration;
- To affect organizational change and accommodate information not hitherto available;
- To reduce maintenance costs;
- To achieve efficiency;
- To improve quality;
- To reduce system complexity;
- To achieve cost effectiveness.

One crucial factor involved with reengineering is to decide when to reengineer. Basically, the decision to reengineer rests on the balancing of cost and risk with tangible benefits. The four factors shown in Figure 15.1 must be considered very carefully in making a reengineering decision.

The risk is concerned with determining the degree of uncertainty involved in reengineering. The cost is concerned with comparing the reengineering cost to the cost of maintaining, redeveloping, and reinventing the existing system. The time is concerned with determining the time factor associated with the reengineering effort and the remaining life of the system.

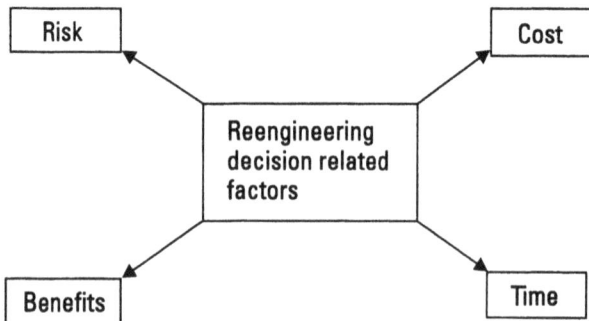

Figure 15.1 Factors to be considered in making a reengineering decision.

The benefits are concerned with determining tangible and intangible advantages of redevelopment.

15.5 Product, Process, and Systems Management Level Reengineering

From the point of view of a product, reengineering may simply be described as some form of reworking or retrofitting of an existing product, which may very well be maintenance or refurbishment [17]. Maintenance can be viewed from various different perspectives: proactive or perfective, reactive or corrective, interactive or adaptive [18]. Alternatively, reengineering may also be viewed as reverse engineering in which the characteristics of the already existing product are identified in order to modify or reuse the product.

In this case, reengineering is product focused; thus, we call it product reengineering. There are many synonyms for product reengineering, including redevelopment, modernization, retrofit, repair, and renewal.

The concept of reengineering may also be applied at the process level, where it is basically concerned with modifications to the current standards cycles in use within an organization in order to benefit from new and emerging technologies or simply to satisfy new requirements of a customer with respect to an item, product, or system. Consequently, process reengineering may be defined as the examination, study, and modification of the functionality or the internal mechanisms of an existing process/system engineering life cycle, for the purpose of reconstituting it in a modern form with modern functional/nonfunctional features without changing the fundamental objective of the process under consideration.

Sometime ago, AT&T Bell Laboratories practiced a similar approach to process reengineering and reported benefits such as [19]:

- Shorter development cycles;
- Less engineering change orders;
- Satisfied customers with respect to their expectations concerning products under consideration;
- Lower program or product development costs throughout the life cycle.

Reengineering at the level of systems management is mainly concerned with potential changes to the organizational processes or entire business, in addition to the systems acquisition process life cycle. Consequently, in a similar manner to process reengineering, systems management reengineering may be defined as the examination, study, and modification of functionality or internal mechanisms of existing system management processes or procedures in an organization for the purpose of reconstituting them in a new form and with modern features, without any changes to the overall organizational objective [19]. There are basically three types of companies that attempt reengineering [6]:

- Type I: Ambitious and seeking to avoid impending difficulties;
- Type II: Passing through rather a difficult period;
- Type III: Anticipating difficulties down the pipeline.

15.6   Reengineering Process

The process of reengineering is composed of six steps, as shown in Figure 15.2. These steps are to plan for reengineering, conduct a feasibility study, get started, perform analysis, design, and implement and monitor. Each of these steps is described next.

Plan for reengineering.   A plan may simply be described as a statement of overall business philosophy, and it presents a sense of high-level direction. In the case of reengineering, it must highlight concerned critical processes. A process in a business is a set of activities that must be conducted successfully

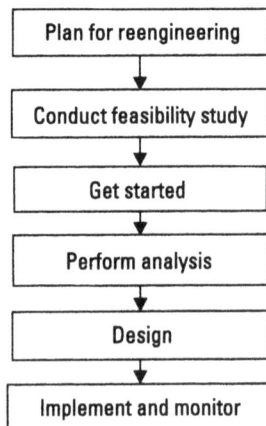

Figure 15.2 Reengineering process steps.

to achieve a business goal. When planning for reengineering, one is concerned with factors such as aligning corporate goals, objectives, vision, and corporate strategy; stating associated constraints; getting senior management commitment; and establishing the final plan for reengineering.

Conduct feasibility study. This is an important step and is basically concerned with determining the feasibility of the reengineering project. More specifically, feasibility is concerned with factors such as weighing risks against doing nothing and determining tangible and intangible benefits.

Get started. Just as in the case of most projects, a team has to be formed to start reengineering. In team composition, careful consideration must be given to the balance between the idealists and the pragmatists, the quantitatively and qualitatively oriented, and the efficiency experts and the humanists. All in all, the reengineering team must be a good representative of organization's vital functions.

Perform analysis. This is concerned with a detailed analysis of the project specifications. This activity is also known as the *diagnostic phase.* The two important tasks involved are stating specifications in operational terms and providing specifications as a basis for design.

Design. In reengineering, design can be due to a fundamental analysis of the organization in question, and a redesign can be due to an analysis of organization structure, reward structure, job definitions, control processes, and business work flows [20]. Design involves tasks such as examining alternatives through performing simulation, prototyping, and best-case studies; making fundamental analysis of business; and doing reverse engineering as appropriate.

Implement and monitor. In this step, the implementation and monitoring of reengineering occur. Implementation requires the configuration of equipment, tools, and human resources. The deficiencies realized through monitoring are fed back to the analysis and design phases for correction.

15.7 Reengineering Tools and Methodology

It is important that reengineering practitioners must choose appropriate tools and approaches that satisfy their requirements most effectively. In so doing, they should seek information on areas such as those listed here [21]:

- The type of reengineering project being undertaken;
- The scope of the project;
- Role consultants (if any);
- People likely to be on the reengineering team;
- Expectations of management with respect to reengineering.

15.7.1 Reengineering Tools

Usually, reengineering tools are utilized more often on projects based on a methodology than on intuitiveness. In fact, the basis for some methodologies is the use of particular tools.

Some of the expectations of the reengineering practitioners in using reengineering tools are shown in Figure 15.3.

In order to produce the benefits shown in Figure 15.3, the reengineering tools should possess characteristics such as ease of use by businesspeople; enhancement of the clarity of the reengineering team's vision; enforcement of the consistency in analysis and design; allowance of interactive, top-down refinement from the reengineering project objectives to the solution; and generation of an acceptable return on investment. Reengineering tools can be classified into six distinct categories as follows [21]:

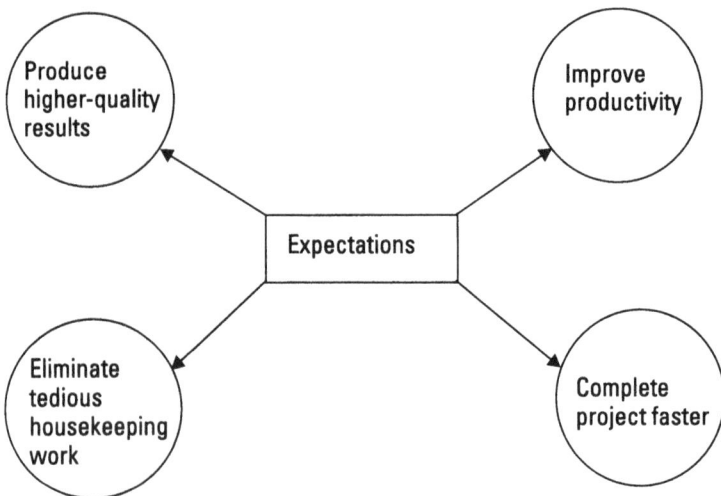

Figure 15.3 Reengineering practitioners' important expectations from using reengineering tools.

1. *Coordination.* The purposes of the tools belonging to this category are to distribute plans and to communicate updated project details. The basic subclassifications include bulletin boards, e-mail, shared spreadsheets, and scheduling applications.

2. *Modeling.* The basic purpose of the tools belonging to this category is to develop a model of something in order to understand its workings and structure. The majority of tools that fall under this classification are integrated computer-aided software engineering tool sets for integrated analysis, design, and development of computer systems. These include items such as Popkin System Architect, S/Cubed DAISYS, Knowledge Ware's IEW, and Texas Instruments' IEF.

3. *Business-process analysis.* The purpose of the tools belonging to this category is to systematically reduce a business into its constituent parts and examine interactions among those parts. Broadly speaking, the tools used under the modeling category can also be used for business-process analysis.

4. *Systems development.* The purpose of the tools belonging to this category is to automate the reengineered business processes. Examples of the tools are ICASE tools, application frameworks (e.g., Borland Application Framework, Gupta SQL, Base, and SQL Windows), object reuse libraries (e.g., Digitalk Smalltalk V, Borland Object Vision), test harness (e.g., McCabe & Associates Code breaker, Software Research M-Test), coding workbenches (e.g., Micro Focus COBOL/2 Workbench, IBM OS/2 Work frame 2), and visual programming (e.g., Microsoft Visual Basic).

5. *Project management.* The purposes of the tools belonging to this category are to plan, schedule, budget, track, and report reengineering projects. Some examples of these tools are Harvard Project Manager, Microsoft Project for Windows, and Texas Instruments' IEF/Project Manager.

6. *Human resources analysis and design.* The purposes of the tools belonging to this category are to design and establish the human or social part of reengineered processes. One subgroup of these tools (e.g., Revelation HR-Applicant Track and Spectrum HR: AM/2000) is utilized for purposes such as requisition or candidate tracking and position history. Other subgroups include organization charting (e.g., Harvard Graphics and Corel Draw), team building (e.g., Supersynch), skills assessment (e.g., Performance

Mentor), and compensation planning (e.g., Hi-Tech Employee Evaluation and Salary Manager).

15.7.2 A Business Process–Reengineering Methodology

This methodology is quite similar to the reengineering process discussed earlier and is being taught in American Management Association Seminars [21]. It is composed of five stages containing 54 tasks. Some of the attractive features of the methodology are that it is easy to learn, has low associated overheads, does not require a specific consulting involvement, requires very few tools, and can be used with any or all of the categories of business process–reengineering tools.

The five stages of the methodology are as follows [21]:

1. *Preparation.* This is concerned with organizing, mobilizing, and energizing the people for performing the reengineering project. Four tasks are associated with this stage: recognize need, train team, execute workshop, and plan change.

2. *Identification.* This is concerned with developing a customer-oriented process model of the business. Nine specific tasks are associated with this stage: model customers, define entities, model processes, define and measure performance, map organization, extend process model, map resources, prioritize processes, and identify activities.

3. *Vision.* This is concerned with choosing the processes to reengineer and formulate redesign options that can achieve breakthrough performance. Ten specific tasks are associated with this stage: understand process flow, determine performance drivers, estimate opportunity, understand process structure, integrate vision, identify value-adding activities, define subdivisions, benchmark performance, envision the ideal (internal), and envision the ideal (external).

4. *Solution.* This is concerned with defining the technical and social requirements for the new processes in addition to developing implementation plans in detail. This stage is composed of 12 social and 10 technical design-related tasks as presented in Tables 15.1 and 15.2, respectively.

5. *Transformation.* This is concerned with implementing the reengineering plans. Nine specific tasks are associated with this stage:

Table 15.1
Social Design-Related Tasks

Number	Task
1	Define team and jobs.
2	Empower customer-contact personnel.
3	Design appropriate incentives.
4	Design appropriate career paths.
5	Specify job changes.
6	Define skills and staffing requirements.
7	Redraw appropriate organizational boundaries.
8	Design change-management program as appropriate.
9	Highlight job-characteristic clusters.
10	Specify management structure.
11	Plan implementation.
12	Define transitional organization.

Table 15.2
Technical Design-Related Tasks

Number	Task
1	Modularize as considered appropriate.
2	Apply technology.
3	Redefine possible alternatives.
4	Instrument and inform.
5	Model entity relationships.
6	Reexamine all possible process linkages.
7	Plan for an effective implementation.
8	Specify deployment.
9	Consolidate interfaces and information.
10	Relocate and retime controls as appropriate.

conduct technical design, train staff, construct system, complete business system plan, pilot new process, improve continuously, refine and transition, develop test and rollout plans, and evaluate involved manpower.

15.8   Reengineering Team and Manpower

A team of competent individuals is necessary to perform the reengineering work effectively. This section discusses various aspects of the reengineering team and reengineering manpower.

15.8.1   Reengineering Team

This is made up of a group of people with the sole responsibility to reengineer a particular process by conducting analyses of the existing one and overseeing its redesign and implementation. In selecting team members, careful attention must be paid so that they are good team players, open minded, creative, and well regarded among stakeholders, peers, and other business leaders. Nonetheless, the team should include individuals with a high level of expertise in the process under consideration, individuals without having any knowledge in the process, experts in technology, individuals representing impacted organizations, representatives from customers, and individuals from outside the company.

The size of a reengineering team should be from 3 to 12 members [22]. Ideally, these members should spend 100% of their time on the reengineering effort, but it must never be less than 75% [4].

15.8.2   Reengineering Manpower

This section discusses various individuals involved in reengineering.

Group or team leader.   This individual is normally a senior executive who authorizes and motivates the company-wide reengineering effort. The primary role of the reengineering team leader is to act as visionary and motivator in addition to demonstrating leadership through symbols, signals, and systems [4]. More specifically, the functions of a reengineering team leader include planning the project, taking accountability for the project outcome, managing the budget, choosing the methodology, leading the team, and interacting with the steering committee [22].

Process owner or leader.   This individual is usually a senior manager vested with responsibility for a specific process on which the reengineering effort is focused. The process owner or leader ensures that the reengineering effort is accomplished successfully at the process level.

Reengineering expert or czar. This individual is concerned with developing reengineering tools and methods within the company as well as accomplishing synergy across the organization's distinct reengineering efforts. A reengineering expert or czar may simply be called reengineering chief of staff of the group or team leader, with two important functions: enable and support each process owner or leader and reengineering team and coordinate all active reengineering activities.

All in all, a creative and fruitful reengineer should possess qualities such as sensitivity (e.g., diplomatic, nonabrasive), good communication skills (e.g., able to solicit buy-in and ownership), visionary skills (e.g., a futurist), open mindedness (e.g., receptive to the ideas of others), leadership ability (e.g., sets the pace and determines the tone), creativity (e.g., able to formulate outstanding alternatives), accomplished achieving history (e.g., has a good track record), commitment (e.g., has the discipline to see projects to fruition), credibility (e.g., already enjoys liking and respect), and ability to see the big picture (e.g., the forest from the trees) [5].

15.9   Reengineering Guidelines and Product-Reengineering Risks

Some of the guidelines for sustaining and succeeding reengineering in an organization are as follows [10]:

- Choose reengineering team members carefully.
- Accept failure, learn from it, and remain focused.
- Move reengineering projects with lightning speed.
- Tailor reengineering to the characteristics of the organization under consideration.
- Diffuse and leverage learning from each and every reengineering project within the organization.
- Communicate broadly, truthfully, and via multiple forums to all concerned.
- Acknowledge up front that all concerned individuals must participate effectively in a mindset change to enable the success of the reengineering initiative.

- Foster commitment and ownership with respect to reengineering at all organizational levels.

- Exploit *clean slate* opportunities as much as possible.

- Aim to ascend to higher forms of reengineering with the passage of time.

There are many types of risks associated with product reengineering that require careful consideration during the reengineering process. Table 15.3 presents nine types of such risks [18, 23].

The integration risk is concerned with the product under reengineering being unable to be integrated effectively with the existing systems. The schedule risk is concerned with schedule delays with respect to the product under reengineering meeting all specifications. The process risk is concerned with the product under reengineering possibly improving situations despite a specific defective organizational process in which the product in question is to be used. The tool and method availability risk is concerned with the reengineering of the product on the assumption that the required approaches will be within reach. The cost risk is concerned with the product under reengineering meeting specifications only after major cost overruns. The maintenance-improvement risk is concerned with the product under reengineering raising rather than lowering maintenance-related problems.

Table 15.3
Types of Product Reengineering–Related Risks

Number	Risk Type
1	Integration risk
2	Schedule risk
3	Process risk
4	Tool and method availability risk
5	Cost risk
6	Maintenance-improvement risk
7	Leadership, strategy, and culture risk
8	Systems-management risk
9	Human-acceptance risk

The leadership, strategy, and culture risk is concerned with the product under reengineering imposing a technological fix on the organizational environment that may not adapt to the product level. The systems management risk is concerned with the product under reengineering imposing a technological fix on a situation that needs organizational reengineering at the systems management level. Finally, the human-acceptance risk is concerned with the product under reengineering being unsuitable for human interactions.

15.10 Reengineering Assumptions, Success and Failure Factors, Top Management Reengineering–Related Mistakes, and Reengineering Sources of Resistance and Myths

A number of assumptions are associated with reengineering. The fundamental assumptions include: Reengineering is top-down directed; reengineering leads to radical change; reengineering focuses on end-to-end processes; reengineering assumes clean-slate change; and reengineering is information technology enabled [13].

As the practice of reengineering does not result in an automatic success, careful consideration in its application is required for ultimate success. Practices such as those listed here are useful for the success of the reengineering application [16, 19]:

- Develop aggressive, but achievable, reengineering performance-related goals with respect to results.

- Obtain top management commitment.

- Aim to assign the task of the reengineering project leadership to a senior management individual.

- Obtain management support at all levels of the organization.

- Secure adequate project resources.

- Focus on opportunities and anticipate resistance.

- Perform analysis of customer needs, market trends, organizational realities, and strategic economic issues before starting the reengineering project.

- Project sufficient time for the project;

- Perform a pilot study in addition to prototyping the reengineering effort to obtain results useful for making refinements to areas such as the reengineering process, building enthusiasm, and increasing communications.

- Train reengineering manpower as the needs arise and hold reengineering workshops for management.

Similarly, some of the factors that contribute to the failure of the reengineering effort are as follows [16, 19]:

- Assigning average-performing individuals to the reengineering task;
- Too many concurrent improvement projects under way;
- Using inadequate reengineering training programs;
- Lacking a positive attitude for reengineering;
- Lacking effective communication during the implementation phase;
- Overlooking the achieved results but measuring the reengineering plans and activities;
- Lacking optimism;
- Overlooking innovative ideas for reengineering due to factors such as politics and risk involved;
- Focusing on reducing cost;
- The wrong sponsor.

As mentioned earlier, top management plays a crucial role in the success of reengineering, but some of its biggest mistakes are as follows [24]:

- Does not provide visible support to items such as role-model changes and reinforcing the change with other managers;
- Changes the direction of the project before completion;
- Delays decisions, thus causing slow project progress and low morale;
- Abdicates project ownership to another manager;
- Fails to set clear objectives and boundaries for the project;
- Does not get directly involved with the project;
- Fails to actively engage with the reengineering team;
- Fails to communicate the business-related reasons for the change and the expected outcome to people such as employees and managers;

- Fails to take sufficient time in understanding current business processes and the mechanism of how business-process reengineering functions;
- Underestimates the amount of time and resources required for reengineering;
- Fails to effectively anticipate the impact of reengineering on involved employees;
- Fails to manage the change within the organizational setup in an effective manner.

There are many organizational-related sources of resistance to reengineering. Some of the basic ones include threatened power, narrow focus of change, threatened expertise, negative group inertia, and resource allocation [5].

Some of the most pernicious and persistent myths about reengineering are that reengineering is automation, reengineering is a synonym for downsizing, and reengineering always means decentralization [14].

15.11   Problems

1. Define the following terms:
   - Reengineering;
   - Restructuring;
   - Data reengineering.

2. List at least nine reasons for reengineering.
3. Describe the following two types of reengineering:
   - Process;
   - Product.

4. What are the factors that should be considered in making a reengineering decision?
5. Describe the steps of the reengineering process.
6. Discuss at least four types of reengineering tools.
7. Discuss the following with respect to reengineering manpower:
   - Team leader;
   - Process owner.

8. Describe at least four types of product reengineering–related risks.

9. What are the factors useful for the success of the reengineering application?

10. List at least 10 top management mistakes with respect to reengineering.

References

[1] Clemons, E. K., M. E. Thatcher, and M. C. Row, "Identifying Sources of Re-engineering Failures: A Study of the Behavioral Factors Contributing to Re-engineering Risks," *Journal of Management Information Systems*, Vol. 12, No. 2, 1995, pp. 9–36.

[2] Bashein, B., L. Markus, and P. Riley, "Preconditions for BPR Success," *Information Systems Management*, Vol. 11, No. 1, 1994, pp. 7–13.

[3] Willets, L. G., "Human Resources: First Stop for Re-engineers," in *Enterprise Re-engineering*, L. G. Willet (ed.), Savoy, IL: Re-engineering Resource Center, Coe-Truman Technologies, Inc., 1996, pp. 8–10.

[4] Hammer, M., and J. Champy, *Reengineering the Corporation*, New York: HarperBusiness, 1993.

[5] Loh, M., *Re-engineering at Work*, Aldershot, U.K.: Gower Publishing Limited, 1995.

[6] Hammer, M., "Re-engineering Work: Don't Automate, Obliterate," *Harvard Business Review*, July/August 1990, pp. 104–112.

[7] "Re-engineering Definitions," approved by the Joint Logistics Commanders Computer Resources Group (JLC/CRM), Department of Defense, Washington, D.C., 1992.

[8] "Re-engineering and Reverse Engineering Terminology," Technical Council on Software Engineering, IEEE Computer Society, Washington, D.C., 1991

[9] Hussain, K. M., and D. Hussain, *Information Technology Management*, Oxford, U.K.: Butterworth/Heinemann Co. Ltd., 1997.

[10] Caron, J. R., S. L. Jarvenna, and D. B. Stoddard, "Business Re-engineering at CIGNA Corporation: Experiences and Lessons Learned from the First Five Years," *Management Information Systems Quarterly*, Vol. 18, No. 3, 1994, pp. 120–139.

[11] Maglitta, J., "Glass Act," *Computer World*, Vol. 28, No. 3, 1994, pp. 80–88.

[12] Hall, G., J. Rosenthal, and J. Wade, "How to Make Re-engineering Really Work," *Harvard Business Review*, November/December 1993, pp. 119–131

[13] Stoddard, D. B., S. L. Jarvenpaa, and M. Littlejohn, "The Reality of Business Re-engineering," *California Management Review*, Vol. 38, No. 3, Spring 1996, pp. 57–76.

[14] Manganelli, R. L., and S. P. Raspa, "Why Re-engineering Has Failed," *Management Review*, July 1995, pp. 39–43.

[15] Taylor, S. T., "Point-of-Care Re-engineering," in *Enterprise Re-engineering,* L. G. Willet (ed.), Savoy, IL: Re-engineering Resource Center, Coe-Truman Technologies, Inc., 1996, pp. 9–14.

[16] Bashein, B. J., M. L. Markus, and P. Riley, "Preconditions for BPR Success and How to Prevent Failures," *Information Systems Management,* Vol. 11, No. 2, 1994, pp. 7–13.

[17] Hunt, V. D., *Re-engineering: Leveraging the Power of Integrated Product Development,* Essex Junction, VT: Oliver Wright Publications, 1993.

[18] Dhillon, B. S., *Advanced Design Concepts for Engineers,* Lancaster, PA: Technomic Publishing Company, 1998.

[19] Sage, A. P., "Systems Engineering and Systems Management for Re-engineering," *Journal of the Systems Software,* Vol. 30, 1995, pp. 3–25.

[20] Guha, S., W. J. Kettinger, and T. C. Tang, "Business Process Re-engineering," *Information Systems Management,* Vol. 10, No. 3, 1994, pp. 13–22.

[21] Klein, M. M., "Re-engineering Methodologies and Tools," *Information Systems Management,* Vol. 11, No. 2, 1994, pp. 30–35.

[22] "Selecting the Right Team for Your Project," Business Process Re-engineering Online Learning Center, Prosci Research, Loveland, CO, 2000.

[23] Arnold, R. S., "Common Risks of Re-Engineering," *IEEE Comp. Soc. Rev. Eng. Newslett.,* April 1992, pp. 1–2.

[24] "Encouraging Top Management: The Role of Executive Leadership in Business Process Re-engineering," Business Process Re-engineering Online Learning Center, Prosci Research, Loveland, CO, 2000.

16

Information-Technology Management

16.1 Introduction

Since the first appearance of electronic computers in the 1940s, the computer industry has become a multibillion-dollar business. Each day the need for an individual to have a working knowledge of computers to perform various types of tasks is increasing. Some of the factors behind this need are the Internet, client-server systems, and collaborative computing and networking. More specifically, it may be said that *information technology* (IT) has grown at a rapid pace within 5 decades, and today we are using fourth-generation languages and fifth-generation computer systems. The management of IT is becoming an important factor in the ability of organizations to compete successfully in the global economy. Many forward-thinking organizations have developed IT management goals. For example, the U.S. Department of Defense has developed the following four goals for IT management [1]:

1. *Become a mission partner.* This includes actions such as increasing and promoting information technology interaction with mission, facilitating process improvement, and serving mission information users as customers.

2. *Provide services that effectively satisfy customer information requirements.* This includes actions such as building architecture and performance infrastructure, upgrading technology base, improving IT

management tools, and modernizing and integrating defense-information infrastructure.

3. *Reform IT management processes for improving efficiency and mission contribution.* This includes actions such as upgrading the IT-associated work force, institutionalizing the IT Management Reform Act provisions, and instituting fundamental IT management–reform efforts.

4. *Secure and protect critical-information sources.* This is concerned with building information-assurance frameworks, building information-assurance architecture and support services, and improving acquisition processes and regulations.

This chapter presents various important aspects of IT management.

16.2   IT Management–Related Facts and Figures

Some of the facts and figures, directly or indirectly, related to information technology management are as follows:

- In 1987, almost 40% of all U.S. capital expenditure accounted for information systems (i.e., $97 billion per year) [2, 3].

- In 1979, 35% of IT budgets in the United States was spent on hardware and only 6.2% on software. However, in 1993 the figures were 19% and 16.6%, respectively [4].

- In 2000, the United States ranked third in the world in Internet connectivity as measured by the number of hosts per 1,000 persons, after Finland and Iceland [5, 6].

- Demand for IT and software systems is growing at an annual rate of at least 25% [6].

- In 1996, it was estimated that the top 500 companies in the United States will spend approximately $100 billion on IT [6].

- In 1996, the expenditure on IT in the United States was approximately 44% higher than in 1995 [6].

- In 2000, it was estimated that the percentage of IT expenditure devoted to software in the United States was more than 50% of total dollars [4].

- Various past studies indicate that approximately 20% to 30% of information-system projects are successful from project inception through institutionalization [5].

16.3 IT Manpower and Its Associated Stress, Sources of Conflict Between Corporate Management and IT Professionals, and IT Productive-Time Estimation

Even though computers are in widespread use, they cannot execute specified tasks without the assistance of IT professionals and support people. More specifically, without the help of people such as analysts, programmers, operators, knowledge engineers, database administrators, and clerks, computers alone will be unable to generate the information end users request and require. Some of the newly emerging positions in IT are policy analyst, manager of technology, technology watcher, and chief information officer. Some of the professionals directly or indirectly concerned with IT are as follows [4, 7]:

Systems Analyst [8]

This individual participates in the development, implementation, and maintenance of information systems. Some of the specific functions of a systems analyst are to gather and analyze data; design forms and procedures; test systems; document systems; study problems and decide which procedures, methods, or techniques are most suitable for a computer solution to the problems; and serve as a link between users and information system department staff.

Some of the desirable characteristics of a system analyst include expertise in systems analysis and system design; creativity; working knowledge of hardware, software, operating systems, and databases; good listening skills; effective knowledge of a programming language; ability to think in the abstract; project-management skills; ability to work on a team; knowledge about clients, their business, and industry; ability to function effectively under pressure; ability to teach and train both professionals and nontechnical users; sensitivity to the company's power structure; and receptiveness to different approaches to problem solving, analysis, and design.

Manager of Technology (MOT) or Chief Information Officer (CIO) [9]

Usually an IT department reports to a MOT, CIO, or director of IT. This individual supervises personnel involved in planning, systems development, operations, and support. Some of the important desirable characteristics of a MOT/CIO are presented in Table 16.1.

Programmer

This individual writes and tests the instructions for directing the computer. A programmer's functions include deciding how to solve a given problem, preparing a logic chart, coding instructions in a computer-understandable language, allocate storage, and preparing appropriate documentation.

Database Administrator

This individual is primarily concerned with the coordination and use of data and knowledge stored under the direction of a database-management system. More specifically, the database administrator minimizes the cost of the machinery or hardware involved in data management, disk space, and time to access data. A database administrator's functions may be grouped into the following four areas:

1. Database design;
2. Database operation;
3. Monitoring;
4. Miscellaneous.

Table 16.1
Important Desirable Characteristics of a Manager of Technology or Chief Information Officer

Number	Desirable Characteristic
1	Appropriate management experience in IT or information systems
2	Good knowledge of IT
3	Appropriate level of skills in communication and facilitation
4	Open mindedness
5	Ability to act as a consultant
6	Loyalty to the company
7	Goal oriented, idea oriented, and systems oriented
8	Comfortable being a change agent
9	Ability to take a global view
10	Comfortable in getting not compliments but blame

Database design includes functions such as data compression, content (i.e., create content and reconcile differences), dictionary or directory (i.e., create and maintain), data integrity (i.e., backup and restart or recovery), and data classification.

Some of the main components of the database-operation areas are database maintenance (i.e., integrity—detect losses and repair losses, recovery, access for testing, and dumping), storage (i.e., physical records structure and physical storage device assignments), security or access (i.e., assign passwords, assign key, and modify passwords), and maintain data element dictionary authority (i.e., add and purge).

The monitoring is concerned with factors such as data quality, performance, efficiency, audit, compliance, and cost.

The miscellaneous area includes functions such as liaison with end users and analysts or programmers, consulting on file design, and providing training on database.

Policy Analyst

This individual identifies potential technological changes, particularly in the area of IT, that may influence the policies and operations of the organization.

Problem Analyst

This professional anticipates and identifies problems that can possibly be addressed by the computing people.

Knowledge Engineer

This professional knows how to represent knowledge and write programs in *artificial intelligence* (AI) languages, which can make inferences and provide management and end users with a sufficient level of expertise found in expert systems. More specifically, a knowledge engineer is well versed in AI tools and methods and interacts with and persuades human-domain experts to articulate their expertise in a precise manner.

16.3.1 IT Manpower–Associated Stress

Stress on people in computer-related professions is becoming an important factor in their productivity and other matters. Frequently, stress is associated with reduction in efficiency, hair-trigger tempers, low morale, and job-hopping. One of the important causes for the stress is physical discomfort in a computer environment. The discomfort may be caused by factors such as screen flicker, glare from terminal, poorly designed or positioned equipment, and loud noises from some peripheral devices.

The fast pace of the computer industry is the main cause of other stress factors. This requires IT professionals to keep abreast of advances and adapt to them as quickly and efficiently as possible. Often, this requires learning new procedures and mastering hardware. Thus, *techno stress* results when a violation of the delicate balance between people and computers occurs.

Management plays an instrumental role in a computerized enterprise by identifying stress causes and initiating appropriate measures to eliminate them. Some of the symptoms of stress among people in a work environment could be as follows [4]:

- Headaches;
- Changes in sleeping and eating patterns;
- Rapid breathing;
- Increase in heart rate;
- General tension of body muscles;
- Feeling of fatigue;
- Nervousness;
- Higher or lower energy level than usual;
- Pain or irritation in neck;
- Circular thought processes.

16.3.2 Sources of Conflict Between Corporate Management and IT Professionals

From time to time, various types of conflicts occur between corporate management and IT professionals. For IT professionals to be successful in performing their tasks effectively, the backing of the corporate management rather than a stressful relationship between the two groups is necessary.

Figure 16.1 presents reasons that there may be tension and conflict between corporate management and IT personnel [4, 10].

In the case of dissemination of information, the corporate management instinctively or deliberately is biased toward restricting the distribution of information due to fear of misuse. On the other hand, IT personnel generally favor the wide distribution of valid information to all involved users irrespective of the subsequent consequences.

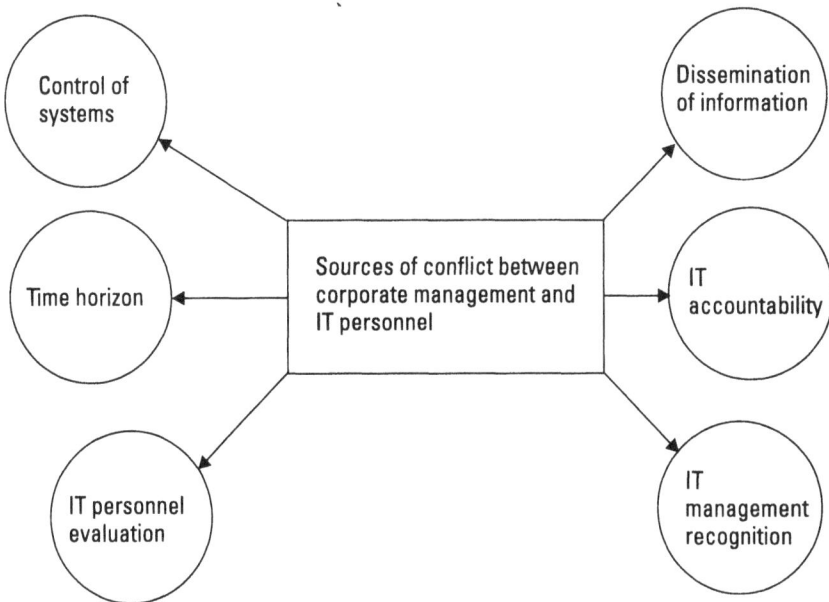

Figure 16.1 Reasons for tension and conflict between corporate management and IT personnel.

The time horizon of corporate management with respect to computer applications is much more short term than for IT people. In fact, management may feel it does not get the output soon enough because IT people plan for the long term and take a holistic and integrative approach. Consequently, a certain degree of unhappiness occurs between these two groups because of the pace of development.

In the case of IT accountability, as many benefits of IT are rather intangible, the corporate management does not always appreciate such benefits as much as IT personnel would like.

In the case of participation in development, IT personnel would like to see effective participation by corporate management in the product development process rather than management simply delegating the responsibility to them.

In the case of IT personnel evaluation, the bone of contention between corporate management and IT personnel is the manner in which corporate management evaluates IT personnel (i.e., use of budgetary, quantitative, and general efficiency measures). IT personnel believe such measures poorly reflect important performance characteristics of IT people.

In the case of IT management recognition, often friction between corporate management and IT personnel or management may be attributed to the belief that corporate executives do not give sufficient recognition or status to IT management. For example, IT management believes that the corporate management rarely consults IT people in making key decisions on products or budgets.

16.3.3 IT Manpower Productive-Time Estimation

This model can be used to estimate productive time of IT manpower and to show that for a large number of people working together in a group, the overall productivity drops. The total productive time is expressed by [4, 11]

$$TPT = nH\left[p - 0.0001\left\{ \frac{n(n-1)}{2} \right\} \right] \qquad (16.1)$$

where

TPT	is the team productive time expressed in hours per week.
p	is the percentage of a calendar week that is considered productive.
H	is the single worker hours per work period.
n	is the number of persons working together in a team.

For $T = 40$ and $p = 55$ in (16.1), Table 16.2 presents values of the *team productive time* (TPT) for the varying value of the team size.

The tabulated values for TPT indicate that as the team size increases, the value of TPT increases, and it peaks at around a team size of 60. It is important to note that for a team size of 20 and one of 94, the value of TPT is roughly the same. Furthermore, the value of TPT is roughly half for a team size of 100 in comparison with a team size of 20.

16.4 Network Management

Networks are growing at an amazing rate and it was predicted that by 2000, there would be 100 million microcomputers tied into corporate networks, a number of them using the Internet. In addition, there are mini and

Table 16.2
Team Productive Time Versus Team Size

Time Size (*n*)	TPT per Week (Hours)
20	424.8
40	755.2
60	895.2
80	748.8
94	424.5
100	220

mainframes that make use of networks daily for transactions such as electronic publishing, reservations, and financial services.

A network may simply be described as a set of nodes connected through links and communications facilities consisting of both physical and logistical elements. Effective management of such networks is essential because losing even a small amount of information vital to a company may result in costing millions of dollars to that organization [12]. Some of the objectives of network management are as follows [4]:

- Improve end users' productivity.
- Detect fraudulent activity.
- Facilitate cooperative work among end users.
- Keep track of the system to highlight potential fault conditions.
- Perform maintenance as required.
- Enforce security around sensitive resources.
- Provide appropriate statistics useful for planning and system control.
- Function effectively without loss of integrity and system security.
- Aim to become robust against misuse and errors.

Figure 16.2 presents five important functions of networks management [4, 13–16].

Problem management is concerned with detecting and fixing problems on a network. Performance management involves monitoring variables

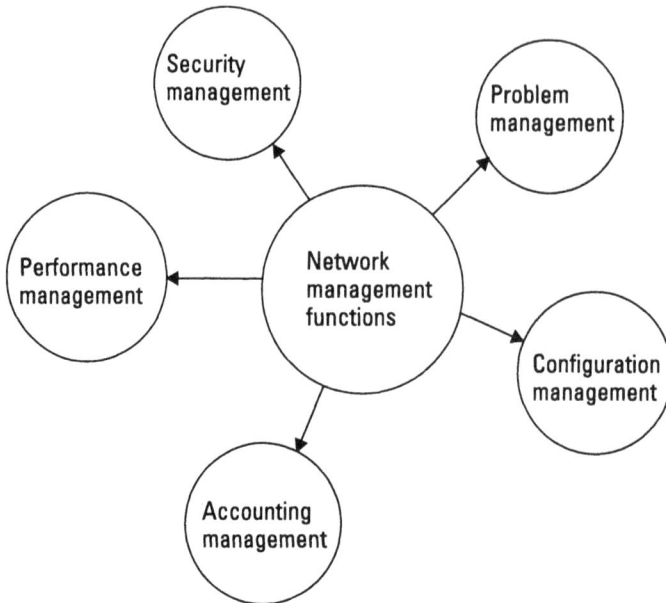

Figure 16.2 Important functions of network management.

relevant to network planning, operation, and maintenance. Some examples of these variables are network availability, traffic density, response time (i.e., the time elapsed from query to output), and utilization of devices.

Accounting management is concerned with the accounting of network assets in the sense of inventory control and tracking and enforcing licenses (if any). Security management involves factors such as detecting and protecting against viruses, controlling access to the network, defining network security zones, erecting and maintaining *firewalls*, and reporting unauthorized use and misuse. Configuration management is basically concerned with initializing and reconfiguring network systems.

16.4.1   Software and Resources for Network Management

In the early days, most functions of network management were performed manually. Nowadays, various software packages are available that can perform almost all of these functions very effectively. Features of such software packages include diagnostic analyzers, virus protection, network traffic monitoring, server monitoring and reporting, alerting, scheduling of tasks, software support, and hardware or software inventory.

Resources required by network management may be classified into two basic categories: hardware and software, and personnel. The hardware and software resources provide information necessary for network management. Most of such information is required for operations at the console during real-time mode. The personnel resource is vital to perform the functions required in network management as well as to manage its resources.

To perform network-management functions, the staff must be properly organized. For the network of a medium-sized firm, the organizational structure may be broken into four major groups: planning and budgeting, design and implementation, operation and maintenance, and administration.

16.5 Client-Server System

This system distributes the load and responsibilities between the processor at the front, known as the client, and the back-end processor, called the server. More specifically, in the client-server mechanism, each and every personal computer is independent for local processing jobs but shares the central equipment (especially servers) through a local-area network [17–21]. Usually, corporate management favors the client-server system, but the professional computing personnel including IT management may not be that enthusiastic about it. The main reason for corporate management to be favoring this system is the reduction in cost. Some of the benefits of the client-server system are as follows [4, 22].

- Allows faster response time in processing requests;
- Reduces responsibilities and overhead cost;
- Provides better access to corporate data and knowledge;
- Provides friendlier interfaces for end users;
- Facilitates greater involvement of end users in the implementation of IT;
- Improves local cost control of development and operations;
- Allows greater possibility for expansion by adding hardware units to networks without replacing the already existing hardware units;
- Provides cooperative processing among individuals and departments across organizational boundaries and geographies;
- Enables distribution of processing from centralized form to desktop computing;

- Provides adequate flexibility in selecting different hardware and network configurations from multiple vendors.

Although the client-server system offers many benefits, there are a number of associated obstacles. These obstacles may be grouped under two different categories: technological and organizational. The technology-related obstacles include inadequate skills and equipment resources, inadequate methodology or experience to plan a client-server system, inadequate development tools, inadequate client-server applications, inadequate national or international standards for the client-server paradigm, and need of local-area network or wide-area network infrastructure.

Similarly, the organization-related obstacles include inadequacy of people skilled in the client-server system and in networking, resistance to change, conversion costs, downsizing risks, and resistance to new technology.

16.6 Development-Information Engineering Approaches, Program Language Selection, and Benefits and Drawbacks of Software Packages

Various types of methodologies and techniques are used by development-information engineering, and some of them are as follows [4, 23]:

- *Jackson structured design.* This approach addresses most of the systems-development life cycle, extending from a general statement of requirements to implementation and maintenance. Furthermore, the method identifies entities as objects and is object-oriented, and its main drawback is the assumption that all vital structures of the problem can be viewed as sequential processes [24].

- *Multiview.* This approach effectively combines structured techniques developed by E. Yourdon [25] with a sociotechnical method, thus emphasizing the social aspects of systems development ignored by most other approaches [26]. The multiview method requires a large gathering of information, and it does explicitly include activities such as training, testing, maintenance, and enhancement but overlooks conversion, cutover, and modification activities.

- *Object-oriented methodology.* This implies the existence of factors such as inheritance, encapsulation, objects, and polymorphism [27].

This methodology can handle complex real-world problems effectively, particularly in interactive environments, and leads to higher programmer productivity and lower maintenance costs of programs. Finally, object-oriented systems have their own design methodology and own programming languages.

- *Merise.* This approach is mainly used in France and is based on the traditional systems life cycle, excluding the planning and implementation phases. The Merise approach addresses conceptual, logical, and physical levels and is described in detail in [28].

- *Structured systems analysis and design method (SSADM).* This approach is restricted to the middle stages of the systems life cycle and has three phases and eight stages as presented in Table 16.3. Reference [29] describes SSADM in detail.

During program language selection, many factors are considered carefully, including program or system objective; internal storage requirements; ease of maintenance; user friendliness; available built-in functions; ability to manipulate data, strings, and graphics; time to compile and execute; technical staff availability; availability of language compiler in 5 to 10 years; available documentation; and special features.

Today, a large number and variety of software packages are available in the open market. These are some of the important benefits and limitations of software packages [4, 30].

Table 16.3
Structured Systems Analysis and Design Method Phases and Stages

Number	Phase	Stages
1	Feasibility study	Problem definition
		Project identification
2	Systems analysis	Analysis of operations and current problems
		Specification of requirements
		Selection of technical options
3	Design	Data design
		Process design
		Physical design

Benefits

- Less costly in comparison to developing customized software;
- Software validated;
- Problems debugged;
- Faster to acquire;
- Documented;
- Frequently periodically upgraded and maintained by vendor;
- State-of-the-art technology.

Limitations

- Usually inappropriate for complex applications;
- Often require tailoring or fine-tuning to satisfy local needs;
- Unavailable for specialized applications;
- Can result in legal problems with respect to copying;
- May require painful or long administrative procedures;
- Require users to conform to format or procedural constraints of the packaged software.

16.7   Human Factors in Information Systems

As human factors play an important role directly or indirectly in IT management, this section presents guidelines that govern all interface designs and designs of input terminals. Thus, the following guidelines are associated with interface designs [31]:

- Aim for as much consistency as possible with respect to terminology, commands, and format.
- Design in dialogues to yield closure.
- Provide users with a sense of control as much as possible.
- Allow frequent users to make use of short cuts such as function keys, abbreviations, and macro facilities as much as feasible.
- Design displays that only impose short-term memory load.

- Design in mechanisms that provide feedback to the user on the effects of actions taken.
- Design in mechanisms that allow easy reversal of actions.
- Offer simple and straightforward error handling.
- Provide defaults as appropriate;
- Design tolerant systems procedures;
- Design effective menus.
- Design messages with care.
- Design user-friendly response times.

Some of the guidelines associated with the design of input terminals are as follows [4]:

- Ensure that the height of the home row keys is 29 1/4 to 31 inches.
- Ensure that the viewing distance is between 17 1/4 and 19 3/4 inches.
- Keep the center of the screen at a position between 10° and 20° below the horizontal plane and the eye height of operator.
- Keep the keyboard at or below elbow height;
- Provide sufficient room for the operator's legs;
- Keep the angle between the upper and lower arms of the operator between 80° and 120°.
- Keep the angle of the operator's wrist no greater than 10°.

16.8 Problems

1. Define the following two terms:
   - Information technology;
   - Information-technology management.
2. What are the desirable characteristics of a systems analyst?
3. What are the important desirable characteristics of a manager of technology?
4. What are the typical functions of a programmer?
5. Discuss the following professionals:

- Knowledge engineer;
- Database administrator.

6. What are the principal reasons for tension and conflict between corporate management and IT personnel?

7. What is network management? Discuss its objectives.

8. What are the advantages of the client-server system?

9. Describe the followings:

- Jackson-structured design;
- Structured systems analysis and design method;
- Merise.

10. What are the advantages and limitations of software packages?

References

[1] Cohen, W. S., "Annual Report to the President and the Congress," Department of Defense, Washington, D.C., 1999, Appendix K.

[2] Davenport, T. H., and J. E. Sbort, "The New Industrial Engineering: Information Technology and Business Process Redesign," *Sloan Management Review*, Summer 1990, pp. 11–27.

[3] "Office Automation: Making It Pay Off," *Business Week*, October 12, 1987, pp. 134–146.

[4] Hussain, K. M., and D. Hussain, *Information Technology Management*, London: Butterworth-Heinemann, 1997.

[5] Liebowitz, J., *Information Technology Management*, Boca Raton, FL: CRC Press, 1999.

[6] *Year 2000 Survey Results*, Pound Ridge, NY: Rubin Systems, Inc., 2000.

[7] Couger, D., and R. A. Zwacki, *Motivating and Managing Computer Personnel*, New York: John Wiley and Sons, 1980.

[8] Fougere, K. T., "Role of the Systems Analysts as Change Agents," *Journal of Systems Management*, Vol. 24, No. 11, 1991, pp. 6–9.

[9] Grover, V., et al., "The Chief Information Officer," *Journal of Management Information Systems*, Vol. 10, 1993, pp. 107–130.

[10] Smith, H. A., and J. D. McKeen, "Computerization and Management: A Study of Conflict and Change," *Information and Management*, Vol. 22, 1992, pp. 53–64.

[11] Adler, F. P., "Relationships Between Organization Size and Efficiency," *Management Science*, Vol. 7, 1960, pp. 80–84.

[12] Derfler, F. J., "An Eye into the LAN," *PC Magazine*, Vol. 12, November 1993, pp. 277–300.

[13] Muller, N. J., "Integrated Network Management," *Information Systems Management,* Vol. 9, 1992, pp. 8–15.

[14] Broadhead, S., "Network Management," *Which Computer?,* Vol. 15, October 1992, pp. 111–125.

[15] Boehm, W., and G. Ullmann, "Network Management," *International Journal of Computer Applications in Technology,* Vol. 4, 1991, pp. 27–34.

[16] Henderson, L. B., and C. S. Pervier, "Managing Network Stations," *IEEE Spectrum,* Vol. 29, April 1992, pp. 55–58.

[17] "Cover Story: Client/Server Computing," *Datamation,* Vol. 37, No. 20, October 1993, pp. 7–24.

[18] Jeffery, B., "Enterprise Client/Server Computing," *Information Systems Management,* Vol. 13, No. 4, 1996, pp. 7–18.

[19] Levis, J., and P. V. Schilling, "Lessons from Three Implementations: Knocking Down the Barriers to Client/Server," *Information Systems Management,* Vol. 11, No. 2, 1994, pp. 15–22.

[20] Semich, J. W., "Can You Orchestrate Client/Server Computing?" *Information Technology,* Vol. 10, No. 16, 1994, pp. 36–43.

[21] Sinha, A., "Client-Server Computing," *Communication of the ACM,* Vol. 35, No. 7, 1992, pp. 77–98.

[22] Schultheis, R. A., and D. B. Bock, "Benefits and Barriers to Client/Server Computing," *Journal of Systems Management,* Vol. 45, No. 2, 1994, pp. 12–15.

[23] Richmond, K., "Information Engineering Methodology: A Tool for Competitive Advantage," *Telematics and Informatics,* Vol. 8, Nos. 1 and 2, 1991, pp. 41–57.

[24] Jackson, M., *System Development,* Englewood Cliffs, NJ: Prentice Hall, 1983.

[25] Yourdon, E., "Reliability of Real-Time Systems," *Modern Data,* Vol. 5, May 1972, pp. 38–52.

[26] Avison, D. E., et al., "Applying Methodologies for Information Systems Development," *Journal of Information Technology,* Vol. 7, 1992, pp. 127–140.

[27] Diaz, O., "Object-Oriented System: A Cross Discipline Overview," *Information and Software Technology,* Vol. 4, No. 1, 1996, pp. 45–57.

[28] Quang, P. T., and C. Chartier-Kastler, *Merise in Practice,* London: Macmillan, 1990.

[29] Downs, E., P. Clare, and I. Coe, *SSADM: Structured Systems Analysis and Design Method,* Englewood Cliffs, NJ: Prentice Hall, 1988.

[30] Sasserath, J. D., "Buying Packaged Software? Caveat Emptor," *Industrial Management and Data Systems,* Vol. 90, No. 2, 1990, pp. 11–13.

[31] Shneiderman, B., *Designing the User Interface: Strategies of Effective Human-Computer Interaction,* Reading, MA: Addison-Wesley, 1992.

17

Software Engineering Management

17.1   Introduction

Software can be considered a product of engineering just like an automobile, telephone, television, aircraft, or any item that requires a high skill level to transform a given raw material into a useful item. Although some people suggest that the term *software engineering* was coined in 1965 [1], it first came into common currency in 1967 when the North Atlantic Treaty Organization Science Committee called for an international meeting on the topic [2, 3].

Today billions of dollars are spent on software annually. For example, in the early 1990s the U.S. Department of Defense alone spent approximately $30 billion annually on software [4], and the size of computer programs has increased from around 100 lines in the early 1950s to multimillion lines in the late 1990s [5, 6]. Factors such as these have played an instrumental role in highlighting the importance of software engineering management. In fact, several U.S. government studies emphasized the importance of software engineering management during the 1980s and early 1990s [7–9].

Software engineering management may simply be described as a system of procedures, technologies, knowledge, and practices that provides the planning, organizing, directing, controlling, and staffing essential to effectively manage software engineering [10]. Software engineering in this case means the practical application of computer science, management, and other sciences for the purpose of analyzing, designing, constructing, and maintaining software and its related documentation.

This chapter presents important aspects of software engineering management.

17.2   Software Facts and Figures

The following facts and figures are directly or indirectly related to software engineering management:

- The United States ranks number one in terms of computers per capita (i.e., approximately 390 per 1,000 persons) and holds about half of the world's computing power [11].

- Software productivity in the United States is declining at a significant rate (e.g., 10% for the period 1994–1995 and 50% for 1995–1996) [11].

- The results of a survey study indicate that software costs are generally large and rapidly increasing throughout the world, with an average yearly growth in the United States of approximately 12% [12].

- As per a Massachusetts Institute of Technology study, over half of software-development projects fail to meet their targets [13]: In the commercial sector and defense industry, the overexpenditures range from an average of 40% to an average of 210%, respectively. Furthermore, schedule overruns range from 90% to 360%, respectively.

- A program size of around 3,200 words consumes an average code-plus-debug time of approximately 178 hours for an individual programmer [14].

- More than 80% of a software product's life is spent in maintenance [15].

- A software product's typical life span is 1 to 3 years in development and 5 to 15 years in use or maintenance [16].

- Software maintenance accounts for between 40% and 90% of software products' life cycle costs [17].

- The annual software industry in the United States is worth at least $300 billion [18].

- A study involving 400 professional software managers indicates that roughly one in four projects comes in at a cost "reasonably close" to the estimate [19].

- Over 70% of the companies involved in software development develop their software through ad hoc and unpredictable approaches [20].

17.3 Software Engineering Tasks and Reasons for Software Management to Be Different

There are many tasks related to software engineering that require management input to varying degrees. Many of those tasks are as follows [21]:

- Problem analysis;
- Requirements determination;
- Software design;
- Software solution coding;
- Code testing and integration;
- Software installation;
- Software quality assurance;
- Software documentation;
- Project management;
- Resource estimation and training.

There are many factors that make software management different from the management of other engineering fields. Some of those factors are presented next [22].

- Software products are often more complex than other engineering products.

- The software discipline, unlike other engineering fields, does not rest on a sound foundation of physical principles.

- There are not many software management or other professionals with a sufficient degree of experience to fully appreciate the benefits of an effective process because the software engineering field is relatively new.

- Often software is the critical element that couples all the other system elements together.

- The low cost of reproducing software forces software solutions to many late-found system-related problems. Consequently, the software task becomes more complicated.

- Software is most visible, most subject to complaint, and most exposed to requirements changes because it is frequently the system component that presents the function to the end user.

- In other engineering disciplines, release to manufacturing provides a natural discipline—unfortunately, this is not the case with software.

- Often, nonsoftware practitioners view software as a black art. Consequently, many management professionals back away from software issues and do not use their management instincts to find solutions to software-related problems.

17.4   Software Engineering Project Management

Project management may simply be described as a system of management procedures, practices, technologies, skill, and experience that is required to manage an engineering project effectively. When the project product is software, the act of managing the project is referred to as software engineering project management [20].

Often, the software engineering projects are a component of larger projects that include elements such as equipment (hardware), manpower, facilities, procedures, and software. Some examples of the larger projects are aircraft systems, railroad switching systems, and accounting systems. Projects such as these are managed by individuals known as system project managers or program managers. Under their supervision are people such as engineers, scientific specialists, programmers, experts in the field of application, and support staff.

As managers perform the functions of planning, organizing, staffing, directing, and controlling to a varying degree, this section discusses such items with respect to software projects.

17.4.1   Software Engineering Project Planning

The existence of a good plan is essential for every software engineering project. The planning of a software engineering project is composed of factors such as the management activities that lead to selection among alternatives, future courses of project-related actions, and a program or plan for accomplishing those actions. Nonetheless, the software project managers must accomplish the following activities in planning their projects [20]:

- *Develop objectives and goals.* This is basically concerned with determining what the software project must accomplish, when it must be accomplished, and what type and amount of resources are required.

- *Develop strategies.* These strategies may simply be defined as long-range goals and the techniques or mechanisms to achieve those goals.

- *Establish software project–associated policies.* These policies are simply predetermined management decisions.

- *Forecast potential project-related conditions.* This is the responsibility of the project manager to forecast conditions that might impact the software project under consideration. Forecasting involves predicting the future project environment and how the project will respond to the predicted events.

- *Assess risks.* This is concerned with anticipating possible adverse events and problems areas and developing contingency plans, in addition to predicting results of possible courses of action.

- *Determine all possible actions.* This is concerned with development analysis and evaluation of distinct ways to execute the project.

- *Make planning-related decisions.* This is concerned with reviewing and selecting a most suitable action out of various alternatives.

- *Establish procedures and rules.* This is concerned with establishing approaches, guides, and limits for accomplishing the project activity under consideration.

- *Develop software project plans.* This is concerned with specifying all of the actions appropriate to effectively deliver a software item. Typically, the software plan specifies items such as policies, tasks, procedures, rules, schedules, and resources.

- *Prepare budget.* This is concerned with placing cost figures on the project plan and allocating estimated costs to project functions, activities, and tasks.

- *Document the final plan.* This is concerned with formally documenting policy decisions, budget, program plans, courses of action, and contingency plans.

17.4.2 Software Engineering Project Organizing

Organizing a software engineering project involves factors such as developing an effective organizational structure for assigning and accomplishing project-related tasks and establishing the appropriate authority and responsibility relationships among the tasks. Figure 17.1 presents activities that must be accomplished by the project manager when organizing a software project [20]. These activities are: identify and group project-related decisions, choose organizational structures, create organizational positions, establish authority and responsibilities, define position qualifications, and document decisions.

In the case of identify and group project-related decisions, the manager reviews project requirements, defines the tasks to be accomplished, and sizes and groups the involved tasks. Choosing organizational structures calls for selecting the most effective structure to not only complete the software project, but to monitor, control, communicate, and coordinate as well. Creating organizational positions is concerned with establishing titles, job descriptions, and job relationships for each expected project role. In the case of establishing authority and responsibilities, the software project manager defines and assigns the responsibilities and authorities to each organizational position within the framework of the project.

Defining position qualifications calls for identifying qualifications for each position in the project by considering factors such as the appropriate level of experience required, the type of individuals and educational level

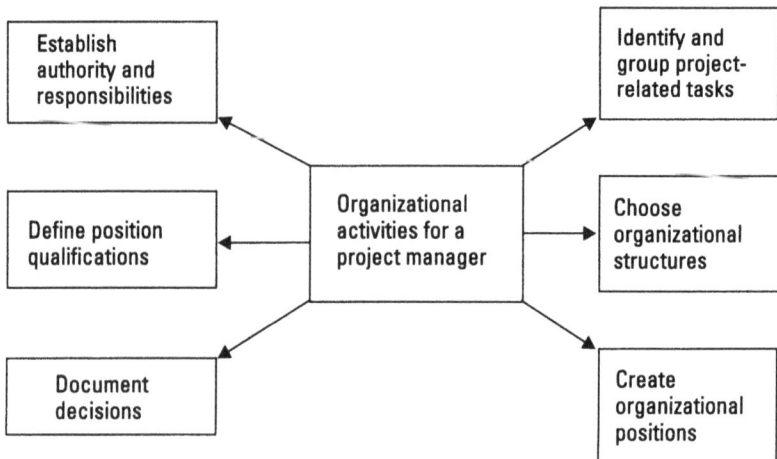

Figure 17.1 Activities to be accomplished by the project manager when organizing a software project.

required, and the need for training. The document decisions are concerned with documenting items such as titles, positions, jobs and their associated descriptions, responsibilities, authorities, relationships, and position qualifications.

17.4.3 Software Engineering Project Staffing

The basic objective of the staffing function is to ensure that software project–related roles are filled by well-qualified individuals. The staffing activities include filling organizational positions, assimilating newly assigned personnel, educating or training personnel; improving knowledge, skills, and attitudes of project-associated personnel; evaluating and appraising manpower, compensating staff; terminating assignments as necessary, and documenting staffing-related decisions.

17.4.4 Software Engineering Project Directing

This is the motivational and interpersonal management aspect that encourages the individuals involved with the project to understand and contribute to project objectives or goals. Furthermore, after the training and orientation of the people involved with the project, the project manager has a continuing responsibility to guide them toward better performance, clarify their assignments, and motivate them to perform assigned tasks with enthusiasm and confidence. Managers perform the following specific activities when directing [20]:

- *Provide leadership.* This is concerned with creating an environment in which project members can accomplish their assigned tasks with enthusiasm and confidence.

- *Supervise involved personnel.* This is concerned with providing day-to-day instructions so that involved individuals accomplish their assigned tasks effectively.

- *Delegate authority to appropriate individuals.* This is concerned with allowing project-associated personnel to make decisions and expend resources within the boundaries of their roles.

- *Motivate.* This is concerned with creating a work environment in which individuals involved with the software project can fulfill their psychological needs effectively.

- *Build teams.* This is concerned with creating a work environment in which involved individuals can work together in harmony to successfully reach the common project goals.

- *Coordinate project activities.* This is concerned with combining software project activities into effective and efficient arrangements.

- *Facilitate communication.* This is concerned with ensuring a free flow of accurate information among entities involved with the software project.

- *Resolve conflicts.* This is concerned with encouraging constructive differences of opinion and cultivating an ideal environment to resolve the resulting conflicts.

- *Manage changes.* This is concerned with stimulating creativity and innovation to achieve project goals.

- *Document directing-related decisions.* This is concerned with documenting decisions that involve delegation of authority, change management, conflict resolution, and communication and coordination.

17.4.5 Software Engineering Project Controlling

This is the collection and analysis of management activities employed for ensuring that the software project goes as per the plans. More specifically, performance and results are compared with the plans, deviations are noted, and corrective measures are initiated for ensuring conformance of plans and actual events. Control may simply be described as a feedback system that provides information on project progress and asks questions on areas such as schedule, cost, and potential problems. The software project–controlling activities are as follows [20]:

- *Develop performance standards.* This is concerned with developing goals that will be achieved when tasks are accomplished correctly.

- *Develop effective monitoring and reporting systems.* This is concerned with establishing software project–monitoring methods and approaches as well as the methods of reporting project status.

- *Measure and analyze results.* This is concerned with comparing achieved results with standards, goals, and plans.

- *Take corrective measures.* This is concerned with initiating corrective measures to bring requirements, plans, and actual project status into conformance.

- *Reward and discipline.* This is concerned with praising, remunerating, and disciplining individuals associated with the software project.
- *Document controlling methods and procedures.* This is concerned with documenting performance standards, control and monitoring systems, and reward and discipline approaches.

17.5 Useful Models for Software Engineering Management

Over the years, many mathematical models that can be useful to software engineering management directly or indirectly have been developed. This section presents some of those models.

17.5.1 Basic Effort Estimation Model

This is a useful model to estimate the level of basic effort required in a software development project under the following three conditions:

1. *Organic.* This means relatively small software teams develop familiar software in an in-house type of environment and the majority of individuals associated with the project possess earlier experience working with similar systems within the company.
2. *Embedded.* This means the software project under consideration may require new technology, unfamiliar algorithms, or an innovative approach to find the solution to a problem. Probably the most important characteristic of the embedded type is the requirement to function within rather tight constraints.
3. *Semidetached.* This is simply an intermediate stage between the organic and embedded types.

For these three conditions, the level of basic effort can be estimated by using the following equation [23]:

$$PM = A(KDSI)^x \qquad (17.1)$$

where

PM is person-month.

$KDSI$ is the delivered source instructions expressed in thousands.

A   is a constant whose value varies with the software develop-
           ment type as presented in Table 17.1.

x   is a factor whose value depends on the software develop-
           ment type as presented in Table 17.1.

17.5.2 Basic Development Time Estimation Model I

This mathematical model is quite useful to estimate software development
time under three different conditions: organic, embedded, and semidetached. Thus, the development time for a software project can be estimated
by using the following equation [23]:

$$DT = B(PM)^y \qquad (17.2)$$

where

DT   is the software development time expressed in months.

B   is a constant whose value is given in Table 17.2.

y   is a factor whose value depends on the type of software
           development project as given in Table 17.2.

Example 17.1

Assume that a software development project is of the embedded type and its
estimated size is 64,000 lines of code. Estimate the number of *full-time-
equivalent software personnel* (FSP) required.
    By substituting the given data into (17.1), we get

$$PM = 3.6\left(64\right)^{1.20} \approx 530 \text{ person-months}$$

Table 17.1

Values for *A* and *x* for Different Software Development Types

Number	Software Development Type	Value of *A*	Value of *x*
1	Organic	2.4	1.05
2	Embedded	3.6	1.20
3	Semidetached	3	1.12

Table 17.2
Values for *B* and *y* for Different Types of Software Development Projects

Number	Software Development Project Type	Value of *B*	Value of *y*
1	Organic	2.5	0.38
2	Embedded	2.5	0.32
3	Semidetached	2.5	0.35

Using the above value in (17.2), we get

$$DT = 2.5 (530)^{0.32} \approx 19 \text{ months}$$

Thus, the number of FSP is

$$= \frac{530}{19} \approx 28$$

It means the project will require approximately 28 full-time-equivalent software personnel.

17.5.3 Full-Time-Equivalent Software Maintenance Personnel Estimation Model

Software maintenance is very important because it represents a large proportion of software life cycle cost. Software maintenance may simply be described as the modification of already existing software while leaving its basic functions intact. In the estimation of *full-time-equivalent software maintenance personnel* (FSMP), the following two factors play an important role [23]:

- Annual change traffic, which is the fraction of the software item's source instructions that undergoes change during a given year.
- Annual maintenance-related effort expressed in person-months.

Annual change traffic is expressed by

$$ACT = (ADSI + MDSI) / TDSI \qquad (17.3)$$

where

ADSI   is the number of added delivered source instructions.

MDSI   is the number of modified delivered source instructions.

TDSI   is the total number of delivered source instructions in a software development project.

Annual maintenance related effort, AMPM, is given by

$$AMPM = (ACT)(PM) \qquad (17.4)$$

Consequently, we have

$$FSMP = (AMPM) / 12 \qquad (17.5)$$

Example 17.2

Assume that in Example 17.1, 6,000 lines of code were added to the original project and 2,000 were modified during first year of maintenance. Calculate the number of FSPM required to maintain the software.

By substituting the specified data into (17.3), we get

$$ACT = (6{,}000 + 2{,}000) / 64{,}000$$
$$= 0.125$$

By inserting the above result and the result for PM as given in Example 17.1 in (17.4), we get

$$AMPM = (0.125)(530)$$
$$= 66.25 \, PM$$

Using the above value in (17.5) yields

$$FSMP = \frac{66.25}{12}$$
$$\approx 6 \, FSMP$$

It means approximately six full-time-equivalent software personnel will be required to maintain the software.

17.5.4 Software Product Development Time Estimation Model II

This is a useful mathematical model to estimate the time required to develop a software product when its size is known. The length of time required to develop a software product is expressed by [24]

$$T_{sd} = \left[\frac{SS}{(k)(DE)^{\alpha}} \right]^{1/\beta} \tag{17.6}$$

where

T_{sd} is the time required to develop a software product, expressed in years.

SS is the software product size expressed in source lines of code.

k is the technology constant, and it takes into consideration the complexity of the software to be developed and the development environment sophistication. More specifically, k is a generalized productivity measure.

DE is the development effort expressed in years.

α, β are parameters and their values are established on the basis of the organization's experience. In the event of having no such values, their recommended values are 0.6288 and 0.5355, respectively.

Example 17.3

Assume we have $SS = 200,000$; $k = 4,000$; $\alpha = 0.6288$; $\beta = 0.5355$; and $DE = 166.7$ labor years (i.e., equivalent to a development productivity of 150 SLOC per labor-month).

Estimate the value of T_{sd} by using the specified data in (17.6). By substituting the given values into (17.6), we get

$$T_{sd} = \left[\frac{200,000}{(4,000)(166.7)^{0.6288}} \right]^{\frac{1}{0.5355}}$$

$$= 3.66 \text{ years}$$

It means the time required to develop the software product under consideration will be 3.66 years.

17.5.5   Software Maintenance Cost-Estimation Model

This model can be used to estimate software maintenance cost. The software maintenance cost is expressed by [25].

$$SMC = 3 \left(C_{mm} \right) \theta / DC \qquad (17.7)$$

where

SMC	is the software maintenance cost.
DC	is the difficulty constant and its value depends on the type of program: easy (500), medium (250), and hard (100).
C_{mm}	is the cost per man-month.
θ	is the total number of instructions to be modified per month.

17.5.6   Source-Code Readability Index

This index is useful in measuring source-code readability and is specifically designed for software products. The index is defined by [26].

$$I_{scr} = 0.295\,L - 0.499\,M + 0.13N \qquad (17.8)$$

where

I_{scr}	is the source-code readability index.
N	is McCabe's complexity or cyclomatic number.
M	is the number of lines containing statements.
L	is the mean normalized length of variables. A variable's length is the total number of characters in a variable name.

17.5.7   Software Size-Estimation Model

This is a useful model to estimate software size by analogy. More specifically, the model estimates the size of new software by using the size of previously developed software for the similar application. Thus, the size of the new program is expressed by [27, 28]:

$$SNP = \left[\frac{SAS}{\gamma 0.678} \right]^{p^{0.678}}$$ (17.9)

where

SNP   is the size of the new program expressed in object words.

SAS   is the size of the analogous software expressed in object words.

γ is the total number of major functions carried out by the analogous software.

p is the total number of projected major functions to be carried out by the new program or software under consideration.

17.5.8 Fog Index

As readability affects maintainability, particularly for textual products, this index could be a useful readability measure in software work. The index is expressed by [29–31].

$$FI = 0.4 \left[\frac{TW}{TS} + \lambda \right]$$ (17.10)

where

FI   is the Fog index.

TW   is the total number of words.

TS   is the total number of sentences.

λ is the percentage of words containing three or more syllables.

The value of this index is purported to correspond roughly with the years of education a person will need to easily comprehend a passage.

17.6 Software Engineer's Functions and Skills

Today the software engineer is one of the well-known engineering professionals and performs functions such as those listed here [32].

- Developing and verifying user requirements;
- Examining software-design methods;
- Selecting software-design methods;
- Structuring software tests;
- Specifying software tests;
- Planning software-design reviews;
- Implementing software-design reviews;
- Selecting software tools such as compilers, validators, and test beds;
- Critically reviewing methods at the end of the life cycle.

A software engineer must have skills in a number of areas, and some of the important ones are given in Figure 17.2: test methods (static and dynamic), discrete mathematics, design methods, social issues, and microelectronics.

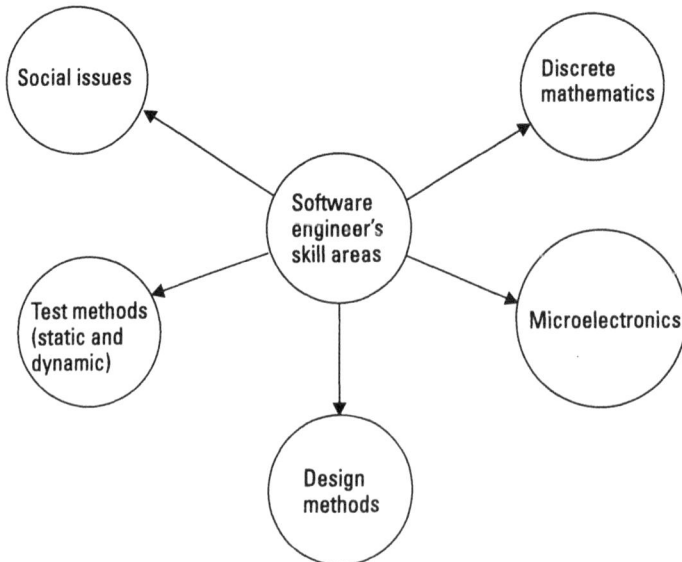

Figure 17.2 Software engineer's skill areas.

Both static and dynamic test methods are concerned with verifying the software system being developed during design. Discrete mathematics is used to tackle more formal specification methods and to utilize formal verification approaches. Design methods are concerned with providing traceability and maintenance information within the software design. Social issues deal with topics such as understanding the safety and liability implications of the impact of total systems. Knowledge in microelectronics becomes useful in system understanding.

17.7 Software Engineering Management Standards

Over the years many software-engineering management-related standards have been developed by organizations such as the IEEE, the American Society for Testing and Materials (ASTM), the Department of Defense (DOD), the European Space Agency (ESA), the Jet Propulsion Laboratory (JPL), and the North Atlantic Treaty Organization (NATO). Some of those standards are as follows [33]:

- IEEE 1058.1–1987, *Standard for Software Project Management Plans,* IEEE, Piscataway, New Jersey, 1987.

- IEEE 982.1–1988, *Standard Dictionary of Measures to Produce Reliable Software,* IEEE, Piscataway, New Jersey, 1988.

- IEEE 982.2–1988, *Guide for the Use of Standard Dictionary of Measures to Produce Reliable Software,* IEEE, Piscataway, New Jersey, 1988.

- IEEE 1209–1992, *Recommended Practice for Evaluation and Selection of CASE Tools,* IEEE, Piscataway, New Jersey, 1992.

- IEEE 1045–1992, *Standard for Software Productivity Metrics,* IEEE, Piscataway, New Jersey, 1992.

- IEEE 1061–1992, *Standard for Software Quality Metrics Methodology,* IEEE, Piscataway, New Jersey, 1992.

- IEEE 1220–1994, *Application and Management of the System Engineering Process,* IEEE, Piscataway, New Jersey, 1994.

- ASTM-E-662-94, *Guide for Developing Computerized Systems,* ASTM, Philadelphia, Pennsylvania, 1994.

- MIL-STD-1521B, *Technical Reviews and Audits for Systems, Equipment, and Computer Programs,* DOD, Washington, D.C., 1985.

- DOD-AFSCP-800-43, *Software Management Indicators*, DoD, Washington, D.C., 1986.
- ESA-PSS-05-08, *Guide to Software Project Management Issue*, ESA, European Space Research and Technology Center, AG Noordwijk, Netherlands.
- JPL-D-4011, *Software Management Planning*, JPL, Pasadena, California, 1988.
- NATO-NAT-PRC-1, *Software Project Management Procedure*, NATO, Brussels, Belgium.
- NATO-NAT-STAN-7, *Software Metric Requirements*, NATO, Brussels, Belgium.

17.8 Problems

1. Describe the following two terms:
    - Software engineering;
    - Software engineering management.
2. List and discuss at least ten major tasks of software engineering.
3. What are the factors that make software management different from the management in other engineering fields?
4. What are the major functions of software engineering project management? Describe at least two in detail.
5. List and discuss at least 10 activities of software engineering project planning.
6. Assume that a software-development project is of the organic type and its estimated size is 128,000 lines of code. Estimate the number of FSP required by using (17.1) and (17.2).
7. Discuss the following two indices:
    - Fog index
    - Source-code readability index
8. What are the major functions of a software engineer?
9. Discuss the areas in which a software engineer should possess skills to a reasonable degree.
10. Discuss the following:
    - Size of the software engineering industry in the United States.
    - A software product's typical life span in development.
    - Proportion of a software product's life spent in maintenance phase.

References

[1] Metropolis, N., J. Howlett, and G. C. Rota (eds.), *A History of Computing in the Twentieth Century: A Collection of Essays*, New York: Academic Press, 1980.

[2] Naur, P., B. Randell, and J. N. Buxton (eds.), *Software Engineering: Concepts and Techniques*, New York: Petrocelli/Charter Books, 1976.

[3] Mahoney, M. S., "The Roots of Software Engineering," *Centrum voor Wiskunde en Informatica (CWI) Quarterly*, No. 3, Vol. 4, 1990, pp. 325–334.

[4] Horowitz, B. M., *Strategic Buying for the Future*, Washington, D.C.: Libey Publishing, 1993.

[5] Bennett, K. H., "Software Maintenance: A Tutorial," in *Software Engineering*, M. Dorfman and R. H. Thayer (eds.), Los Alamitos, CA: IEEE Computer Society Press, 1997, pp. 289–303.

[6] Gibbs, W. W., "Software's Chronic Crisis," *Scientific American*, September 1994, pp. 86–95.

[7] *Strategy for a DOD Software Initiative*, Report, Department of Defense, Washington, D.C., October 1, 1982.

[8] *Report on the Defense Science Board Task Force on Military Software*, Office of the Undersecretary of Defense for Acquisition, Department of Defense, Washington, D.C., September 1987.

[9] *Embedded Computer Systems: Significant Software Problems on C-17 Must Be Addressed*, Report No. GAO/IMTEC-92-48, Gaithersburg, MD, May 1992.

[10] Thayer, R. H. (ed.), *Software Engineering Project Management*, Los Alamitos, CA: IEEE Computer Society Press, 1997.

[11] *Year 2000 Survey Results*, Pound Ridge, NY: Rubin Systems, Inc., 2000.

[12] Boehm, B., and P. Papaccio, "Understanding and Controlling Software Costs," *IEEE Trans. on Software Engineering*, Vol. 14, 1988, pp. 1462–1477.

[13] Cooper, K.G., and T. Mullen, "Swords and Plowshares: The Rework Cycles of Defense and Commercial Software Development Projects," *American Programmer*, Vol. 6, May 1993, pp. 41–51.

[14] Sackman, H., W. J. Erikson, and E. E. Grant, "Exploratory Experimentation Studies Comparing Online and Offline Programming Performance," *Communications of the ACM*, Vol. 11, 1968, pp. 3–11.

[15] Charette, R. N., *Software Engineering Environments*, New York: Intertext Publications, 1986.

[16] Fairley, R.E., *Software Engineering Concepts*, New York: McGraw-Hill, 1985.

[17] Foster, J., "Cost Factors in Software Maintenance," Ph.D. thesis, University of Durham, U.K., 1993.

[18] Hopcroft, J. E., and D. B. Krafft, "Sizing the U.S. Software Industry," *IEEE Spectrum,* December 1987, pp. 58–62.

[19] Lederer, A. L., and J. Prasad, "Nine Management Guidelines for Better Cost Estimating," *Communications of ACM,* February 1992, pp. 51–59.

[20] Thayer, R.H., "Software Engineering Project Management," in *Software Engineering,* M. Dorfman and R. H. Thayer (eds.), Los Alamitos, CA: IEEE Computer Society Press, 1997, pp. 358–371.

[21] Leach, R. J., *Introduction to Software Engineering,* Boca Raton, FL: CRC Press, 2000.

[22] Humphrey, W. S., *Managing the Software Process,* Reading, MA: Addison-Wesley, 1990.

[23] Legg, D. B., "Synopsis of COCOMO," in *Software Engineering Project Management,* R. H. Thayer (ed.), Los Alamitos, CA: IEEE Computer Society, 1997, pp. 230–245.

[24] Gaffney, J. E., "How to Estimate Software Project Schedules," in *Software Engineering Project Management,* R. H. Thayer (ed.), Los Alamitos, CA: IEEE Computer Society, 1997, pp. 257–266

[25] Sheldon, M. R., *Life Cycle Costing: A Better Method of Government Procurement,* Boulder, CO: Westview Press, 1979.

[26] De Yong, G. E., and G. R. Kampen, "Program Factors as Predictors of Program Readability," *Proc. IEEE Computer Software and Applications Conference,* 1979, pp. 668–673.

[27] Eddins-Earles, M., *Factors, Formulas, and Structures for Life Cycle Costing,* Concord, MA, 1981.

[28] Doty Associates, Inc., *Software Cost Estimation Study,* Report No. RADC TR-77-220, Rome Air Development Center, Griffith Air Force Base, Rome, NY, August 1977.

[29] Pfleeger, S. L., *Software Engineering: Theory and Practice,* Upper Saddle River, NJ: Prentice Hall, 1998.

[30] Gunning, R., *The Technique of Clear Writing,* New York: McGraw-Hill, 1968.

[31] Dhillon, B. S., *Engineering Design: A Modern Approach,* Chicago, IL: Irwin, 1996.

[32] Smith, D. J., and K. B. Wood, *Engineering Quality Software,* London: Elsevier Applied Science Publishers Ltd., 1987.

[33] Thayer, R. H., "Software Engineering Standards," in *Software Engineering,* M. Dorfman and R. H. Thayer (eds.), Los Alamitos, CA: IEEE Computer Society Press, 1997, pp. 509–523.

About the Author

Dr. B. S. Dhillon is a professor of engineering management in mechanical engineering at the University of Ottawa. There he has served as chairman/director of the Mechanical Engineering Department and the Engineering Management Program for more than 10 years. Dr. Dhillon has published more than 280 articles on engineering management, reliability, maintainability, and related areas. He is or has been on the editorial boards of six international scientific journals. Dr. Dhillon has written 23 books on engineering management, reliability and maintainability management, safety, human factors, design, and related areas; some of his books have been translated into several languages, including Russian, Chinese, and German. Dr. Dhillon also served as general chairman of two international conferences on reliability and quality control held in Los Angeles and Paris in 1987.

Dr. Dhillon is the recipient of the American Society of Quality Control Austin J. Bonis Reliability Award, the Society of Reliability Engineers' Merit Award, the Gold Medal of Honor from the American Biographical Institute, and the Faculty of Engineering Glinski Award for Excellence in Research. He is a registered professional engineer in Ontario and is listed in *American Men and Women of Science, Men of Achievements, International Dictionary of Biography, Who's Who in International Intellectuals*, and *Who's Who in Technology*.

Dr. Dhillon has served as a consultant to various organizations and bodies and has many years of experience in the industrial sector. At the University of Ottawa, he has been teaching engineering management, reliability, and related areas for more than 22 years. Dr. Dhillon received a B.S. in electrical engineering in 1972 and an M.S. in mechanical engineering in 1973 from the University of Wales. He received a Ph.D. in industrial engineering in 1975 from the University of Windsor.

Index

Planning and Design for High-Tech Web-Based Training,
  David E. Stone and Constance L. Koskinen

*Preparing and Delivering Effective Technical Presentations,
Second Edition,* David Adamy

*Reengineering Yourself and Your Company: From Engineer to
Manager to Leader,* Howard Eisner

*Successful Marketing Strategy for High-Tech Firms, Second
Edition,* Eric Viardot

*Successful Proposal Strategies for Small Businesses: Using
Knowledge Management to Win Government, Private Sector,
and International Contracts, Third Edition,* Robert S. Frey

Systems Engineering Principles and Practice, H. Robert Westerman

Team Development for High-Tech Project Managers,
  James Williams

For further information on these and other Artech House titles,
including previously considered out-of-print books now available
through our In-Print-Forever® (IPF®) program, contact:

Artech House	Artech House
685 Canton Street	46 Gillingham Street
Norwood, MA 02062	London SW1V 1AH UK
Phone: 781-769-9750	Phone: +44 (0)20 7596-8750
Fax: 781-769-6334	Fax: +44 (0)20 7630-0166
e-mail: artech@artechhouse.com	e-mail: artech-uk@artechhouse.com

Find us on the World Wide Web at:
www.artechhouse.com

www.ingramcontent.com/pod-product-compliance
Lightning Source LLC
Chambersburg PA
CBHW050521190326
41458CB00005B/1614